Introduction to Textile Fibres

Introduction to Textile Fibres

H. V. Sreenivasa Murthy

WOODHEAD PUBLISHING INDIA PVT LTD

New Delhi, India

Published by Woodhead Publishing India Pvt. Ltd.
Woodhead Publishing India Pvt. Ltd., 303, Vardaan House, 7/28, Ansari Road, Daryaganj, New Delhi - 110002, India
www.woodheadpublishingindia.com

First published 2015, Woodhead Publishing India Pvt. Ltd.

Reprint 2017, 2018

Reprinted 2020

Woodhead Publishing India Pvt. Ltd. ISBN: 978-9-38505-909-4
Woodhead Publishing Ltd. e-ISBN: 978-9-38505-957-5

Typeset by Third EyeQ Technologies Pvt Ltd, New Delhi
Printed and bound by Replika Press Pvt. Ltd.

Contents

Preface

We use textile in one form or another during our journey from the 'CRADLE TO COFFIN'. Therefore, a thorough knowledge of textile fibres is necessary to prevent mistakes that are too often made in buying, selling and using the textile materials.

Identifying the need for Coaching Associateship of Textile Association – ATA Part – I subject on textile fibres, content of this book was prepared some time during 1970. Initially, it was in the cyclostyle book form. Later in the year 1986, the upgraded version of Textile Fibre was selected by the book publication committee of The Textile Association India (TAI) to print it in book form. Hence, the content was further revised and made simple to understand the intercacies of Textile Fibre. Then, the manuscript was edited by Prof. M. L. Gulrajani IIT Delhi. The technical contents were scrutinized by Dr. V. B. Gupta, Professor, Fibre Science, IIT Delhi, and the book was published by the Textile Association (India) in the year 1987.

This book provided necessary information for the beginners. In many textile institutions it was widely referred by the students and staff for teaching diploma and degree courses. In the year 1989 this book won the coveted "Best Technical Book in Textile" – Century Mills Award through Textile Association (India).

Lots of developments have taken place in the last two to three decades in understanding and utilization of textile fibres. In the year 2007, friends and well wishers recommended this book for further revision. Since then, the work of revision is going on. In the present revised edition, certain topics like Bt-Cotton, organic cotton and coloured cotton, bamboo and soyabean fibres, etc., has been included. Statistical figures are updated. More information of micro fibres and nano fibres, etc., are given; exhaustive details of chemical composition and statement of the most commonly used textile fibres, categories of fibre polymers end-use application of some fibres are given in appendix.

I wish to appreciate the encouragement and assistane of my good friend Mr. B. Purushothama, who took complete responsibility of going through the computed manuscript, and the rough text, editing and helping me to bring out the revised edition. I have no words to express my gratitude for the help rendered by him; otherwise this revised manuscript could not have found the light of the day. Not but the least, I wish to thank the members of my family for motivating me to undertake the revision of the book.

Author will be happy to receive criticism and suggestions for further improvement of this book.

H. V. Sreenivasamurthy

Bengaluru, 1 May 2015

1 Fibre characteristics

A study of the various textile fibres employed by the different nations through ages provides an excellent commentary on the progress of civilization and a fair grasp of the industrial life and economic condition of the people.

Some knowledge of textile fibres is necessary to everyone, because textiles have an important bearing on our daily life.

The history of fibre development has been a strange pattern of trial and errors. Ever since fibres were used for making apparel fabrics, climate seems to have had much to do with fibre utilization. A study of textile fibres will show, for example, why certain fabrics are durable and, therefore, serviceable for specific purposes. It will explain why certain fabrics make cool weaving apparel besides giving an impression of coolness when used for decoration.

The study of textile fibres has become a very intricate and exacting science with many ramifications into the properties, sources, processing and utilization of these fibres.

Before a fibre can be properly selected and its relative economic value and practical utilization can be determined, a careful study of all physical and chemical properties, their terminology, definitions and their behaviour under various conditions are essential.

1.1 Definitions of some important terms

1.1.1 Textile

The word Textile comes from the Latin word, *Textile*, and the French word *Texere* which mean *pertaining to weaving of the woven fabrics, woven or capable of being woven; formed by weaving; that may be woven; a woven fabric.* Hence, the textiles are the materials that can be converted into yarns and fabrics of any nature or character. Technically they must be of spinnable length. Then only they can be made into spun yarns.

The word *Textile* when used as a noun means anything woven or suitable for weaving. Now as already seen it is applied to any fabric or product manufactured from fabrics, filament or yarns, natural or artificial – by means of interlacing, interlooping, felting or stitching the fibre web, for example, thread ropes, knitted fabrics, nets, felts, etc.

Hence this is a very comprehensive term that includes all clothing, carpets, curtains, tapes, upholstery, etc.

The word *Textile* when used as an adjective, means manufacture of raw materials, processing machinery, buildings, personnel used in the organization connected with the technology of their manufacture, which includes the following: Textile Mill, Textile Institute, Textile Engineer, Textile Research, Textile Printing, Textile Designer, Textile Technologist, etc.

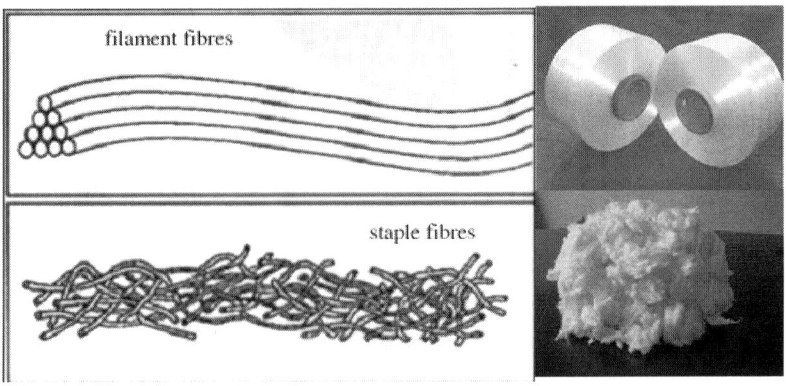

Figure 1.1 Fibre forms

1.1.2 Fibre

In a broad sense the word fibre is used for various types of matter – natural or manmade, forming basic elements of textile fabrics and other textile structures.

It is defined as *one of the delicates, hair-like portion of the tissues of a plant or animal.* Fibre is defined by Fabric Link Textile Dictionary as *The basic entity, either natural or manufactured, which is twisted into yarns, and then used in the production of a fabric.*

The physical interpretation of the word fibre is a unit of matter characterized by having a length of at least hundred times its diameter.

1.1.3 Textile fibre

Technologists have defined the term Textile fibres as those fibres which can be spun into a yarn or made into a fabric by interlacing, or interlooping in a variety of machines including weaving, knitting, braiding, felting, bonding, etc.

1.1.4 Staple

Staple is the name given to the fibres of limited length. To make a continuous length of yarn, staple fibres have to be twisted together. Staple fibres can range from about 1 cm to many centimeters in length. But in no case they ever become long enough to be classified as filament. So the two terms are quite distinct except for the fact that manmade filaments can be converted into staple fibres by deliberately cutting them into short lengths. This is a very common way of processing manmade fibres, but the reverse process is never carried out.

1.1.5 Filament

Filament is the name given to a fibre of continuous length, that is to say, it is long enough to be used in a fabric without increasing its length by adding other fibres to it.

An example of natural filament fibre is silk; the cocoon of a silkworm can contain about 360–1200 m (depending on the quality and type of cocoon) of continuous twin filaments. Manmade filaments produced by spinning machines can be many kilometers long.

There may be one or many filaments in the yarn (see below) and accordingly they are called monofilament or multifilament. The basic difference between a fibre and a filament is that of the length (see Fig. 1.1)

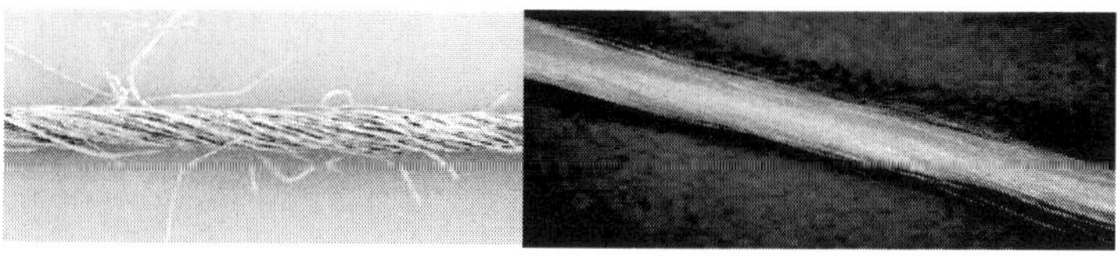

Staple fibre spun yarn

Filament spun yarn

Figure 1.2 Yarn types

1.1.6 Yarn

Yarn is an assemblage of fibres twisted or laid together to form continuous strand suitable for use in weaving, knitting, etc.

Yarn can consist of staple fibres, filaments or combination of the two (Fig. 1.2). Filaments merely need grouping in order to produce the thickness of yarn required, the length being already there in the individual filaments. Grouping of filaments is achieved by twisting them together. The twist, usually quite a small amount, merely serves to keep the filaments reasonably together.

Staple fibres have to be twisted to make them adhere into a continuous length of yarn. The action of twisting, forces the fibre surfaces into contact with each other setting up friction between them which enables them to withstand tension. In this way, a continuous length of yarn can be made even from very short fibres.

This type of yarn exerts a strong influence on the texture and appearance of the fabric. In general, filament yarns are thin, smooth and lustrous, and staple fibre yarns are thicker, fibrous and non-lustrous. An excellent example of these different characteristics can be seen by comparing the fibrous nature of the outer wool fabric of a coat or suit jacket with the smooth lustrous surface of the lining of the garments, the wool fabric being made from staple and lining being made from filament. There are many varieties of yarns such as, single yarns, fancy yarns, cords and twines.

1.1.7 Thread

A thread is three or more yarns tightly twisted, singed, dyed and finished to fit into the eye of the needle, or to be hand-knotted, crocheted or tatted. The very extreme type of thread reaches into the area of cords and eventually ropes of all types.

1.2 Characteristics of a good textile fibre

To be designated as a textile fibre any material should satisfy two important characteristics, namely, the essential or the primary properties and the desirable or the secondary properties.

Essential properties

1. Length
2. Strength and related properties
3. Flexibility
4. Cohesiveness

Desirable properties

1. Fineness
2. Resiliency

3. Uniformity
4. Porosity
5. Lustre
6. Durability
7. Commercial availability

1.2.1 Essential properties of a textile fibre

1.2.1.1 Length

Length is an important parameter which determines the usefulness of a textile fibre from the point of view of spinning. The length of the fibre should not be less than 5 mm. It is of course very easy to understand that when a continuous yarn is to be made out of individual fibres, it should possess a considerable length with reference to its diameter, otherwise it would not be possible to make a yarn that would hold together the constituent fibres. This is referred to as the length to breadth ratio. The most useful fibres should have length to breadth ratios of more than 100:1. Almost all textile fibres have length to breadth ratio of more than 1000:1.

Typical ratios for several natural fibres are as follows:

- Cotton = 1400
- Wool = 3000
- Flax = 1209
- Ramie = 3000
- Silk = 33×10^6

Silk filament has approximately 500 m length and a diameter of around 15 μm. In such case we have the enormous length to breadth ratio of $(500/15) \times 10^6$ or 33×10^6, the silk filament is 33 million times as long as it is broad. These figures have been given in some detail to emphasize that a high length to breadth ratio is the essential characteristic of natural fibres.

In the case of manmade fibres also we find that a high length to breadth ratio is equally important. The diameters to which such fibres are spun are available but in general are of similar magnitude as those of the natural fibres. Thus viscose and cellulose acetate filaments are usually spun with average diameters varying between 10 and 30 μm, which are similar to those found in the natural fibres.

Manmade fibres are mostly used in continuous filament form, and even when they are cut to staple, the staple is seldom less than 25.4 mm in length. It is clear that a high length to diameter ratio is equally a characteristic of manmade fibres as it is of natural fibres.

1.2.1.2 Strength

Strength of any material is derived from the load it supports at break and is thus a measure of its limiting load bearing capacity. Normally strength of a textile fibre is measured in tension when the fibre is loaded along its long axis and is designated as Tensile strength.

Tensile strength of textile fibre is measured as the maximum tensile stress in force per unit cross-sectional area or per unit linear density, at the time of rupture called 'tenacity', expressed in terms of grams per denier or grams per tex units. In Standard International Unit (SI Unit) tenacity is expressed as millinewtons per tex mN/tex (mN/tex = gf/den $\times$ 88.3, mN = gf $\times$ 9.8).

1.2.1.3 Cohesiveness

It is the property of an individual fibre by virtue of which the fibres hold on to one another when the fibres are spun into yarn. This action is usually brought about by the high degree of frictional resistance offered by the surface of the fibres to separate one from the other. The wool fibres, for example, have saw-toothed surface, so that the projecting edges on its surface, called scales, easily catch on to one another when several such fibres are twisted together during spinning. On account of this, fibres offer resistance when an attempt is made to pull them apart. Cotton fibres also possess irregular or rough surface. Further, due to the natural twist in the cotton fibre known as convolution, the fibres interlock themselves by friction when they are spun into yarns. The introduction of crimp in synthetic fibre increases cohesion.

1.2.1.4 Flexibility

The fibre should be sufficiently pliable; then only it can wrap around another fibre during spinning. If, on the other hand, fibre is stiffer and wirier, then it is less adaptable for textile use, for example, glass and metallic fibres.

1.2.2 Desirable properties of textile fibre

1.2.2.1 Fineness

Fineness of a fibre is a relative measure of its size, diameter, linear density or weight per unit length expressed in a variety of units.

Natural fibres vary in fineness more widely than manmade fibres, because in case of manmade fibres the diameter and densities can be controlled very accurately during their manufacture. For example, the variation in the diameter of nylon is only 5–6%, while for natural fibres like wool and silk, it is from 17 to 30%.

Fineness of cotton fibre is expressed in micrograms per inch or per centimeter. For wool fibre, fineness is given in micrometer and for manmade fibres the fineness is give in denier or tex.

- 1 μg = 10^{-6} g;
- 1 μm = 10^{-4} cm = 0.0001 cm. 1 μm = one millionth of a meter or one thousandth of 1 mm, i.e. 0.001 mm;
- 1 Denier = weight in grams of 9000 m;
- 1 Tex = weight in grams of 1000 m).

1.2.2.2 Resilience

The resistance to compression, flexing or torsion varies from fibre to fibre. Some fibres have a natural tendency to return to their original condition when any of the above mentioned forces is applied, an important property where, for instance, recovery from creasing is required. Wool is outstanding in this respect by virtue of its natural characteristics, but cellulosic fibres may be modified in such a manner so as to greatly improve these properties. It is believed that the resilience of a fibre is more a function of molecular cohesion.

This springiness of a fibre or its mass resilience is highly desirable in carpet wool. It is by reason of this quality that wool fabrics hold their shape, drape gracefully and do not wrinkle.

1.2.2.3 Uniformity

Textile fibres should possess uniformity in their thickness and length. Unfortunately none of the principal natural fibres like cotton or wool has the same length and diameter of the fibre in the same lot. Fibres in any specified qualities, grades or lots vary considerably in length and diameter. On the other hand, manmade staple fibres are more uniform as they are cut to the exact length after being spun and drawn, and even the diameter can be controlled within close tolerance limits during its manufacture.

1.2.2.4 Porosity

Porosity can be defined as the ratio of the volume of air contained within the boundaries of the material to the total volume of a solid plus air or void, expressed as percentage.

Porosity facilitates the absorption of moisture, liquid lubricants, dyes, oils and steam by the fibres so as to thoroughly permeate the fibre. Porosity in a fibre is important in wet processing.

The natural and manmade fibres differ greatly in respect of porosity which in turn affects other properties of fibres and consequently the processing of fibres during textile manufacture. In general, natural fibres have higher porosity than synthetic fibres.

1.2.2.5 Porosity lustre

Natural lustre enhances the value of textile fibre especially of natural fibres. For example, the natural lustre of the silk gave it for a long time distinct advantage over the other textile fibres, and experiments were constantly made to improve the lustre of those fibres which were naturally dull. Mercerized cotton, for instance, is a preferred cotton which has not undergone this treatment, owing to the richer appearance of the finished cloth. Since the introduction of viscose, however, with its extremely high and almost metallic lustre, consumer taste has gone a little in the opposite direction, and many fabrics produced today are purposely delustred in order to give the desired matt finish.

Therefore, it is evident that lustre, under certain conditions and for certain purposes, may enhance the value of a fibre. On the other hand, too much of brightness may be a source of aversion to the user and hence it has to be delustred by delustring process down to a required degree of lustre.

1.2.2.6 Durability

A textile fibre should withstand processing treatments and should not be easily susceptible to physical, chemical and bacteriological attack, which may result in damage and decomposition.

The durability of clothing to average wear and tear depends somewhat more on the elasticity, flexibility and resistance of the fibre and fabric, rather than the absolute strength of either fibre or fabric. If a fabric possesses these three properties, its garment will absorb or counter more readily stresses and strains during wear. It will allow itself to be deformed with less resistance, thus reducing the chance of intermediate tearing or twisting. For these reasons wool garments owe much of their durability to the elasticity, resilience and flexibility of the fibre and fabric, even though wool is a weak fibre.

Strength combined with these properties provides excellent durability (see Table 1.1) that is why nylon and polyester fibre fabrics seem to last forever. Strength and reasonable flexibility can also provide good durability as illustrated by cotton which lack elasticity and resilience.

Felting, matting or bonding of fibres in fabrics tends to increase fabric strength and thereby enhance durability. A raised fabric surface increases fabric resilience and provides longer resistance to abrasive surfaces, e.g. carpets, ribbed fabrics, etc.

Table 1.1 Durability evaluations of textile fibres

Fibre		**Rating**
Nylon Polyester Acrylic Viscose Acetate	Manmade fibres	Excellent, outstandingly durable
Cotton Wool Linen Silk	Natural fibres	Poor durability

Source: H.M. Taylor 'Abrasion in Fabrics', *Textiles*, Vol. 7, June 78.

1.2.2.7 Commercial availability

All the essential and desirable properties of a fibre put together will be of much use only if the fibre is available in large quantities at fair price, when it is needed. The accurate estimate of the different type of fibres available for the consumption and the source of availability makes the supply of commercial fibres to establish itself with reasonable assurance of exactness. Following are the properties desired for basic textile fibres:

A. In Apparel and Domestic Application

1. Tenacity: 3–7 grams per denier (gpd)
2. Elongation @ break: 10–35%
3. Recovery from elongation : 100% at strain up to 5%
4. Modulus of elasticity : 30–60 gpd
5. Moisture absorbency: 2–5%
6. Zero strength temperature: Excessive creep and softening point >215 °C
7. High abrasion resistance

8. Dyeable
9. Low flammability
10. Insoluble (low swelling) in water, in moderately strong acids and bases and conventional organic solvents from room temperature to 100 °C
11. Easy care

B. In Industrial applications

1. Tenacity: 7–8 gpd
2. Elongation break: 8–15%
3. Modulus of elasticity: 80 gpd or more, wet: 50 gpd
4. Zero strength temperature: Excessive creep and softening point >250 °C

C. Polymer composition and structure

1. Melting point
2. Modulus
3. Elasticity and recovery from strain
4. Tensile strength
5. Density
6. Moisture absorption
7. Dyeability
8. Comfort
9. Abrasion resistance 1.3

1.3 Classification of textile fibres

Fibres may be classified as either **natural** or **synthetic**.

- The natural fibres may be further classed according to origin as **animal**, **vegetable** or **inorganic fibres**.
- **Animal fibres** are composed chiefly of proteins; they include silk, wool, goat hair (known as mohair), llama and alpaca, vicuña, camel, horse, rabbit, beaver, hog, badger, sable and other animals.
- **Vegetable fibres** are composed chiefly of cellulose and may be classed as short fibres, e.g., cotton and kapok; or long fibres, including flax

(used to produce linen), hemp, Manila hemp, istle, ramie, sisal hemp and Spanish moss.

- The chief natural **inorganic fibre** is asbestos. Fibres are also derived from other inorganic substances that can be drawn into threads, e.g., metals (especially gold and silver).
- **Artificial fibres** can be produced either by the synthesis of polymers (nylon) or by the alteration of natural fibres (rayon).

Classification is branching of whole discipline in different categories. Since textile fibres have ceased to be of descriptive discipline and have become a branch of science, the exactness and intricacy has entered into it. Therefore, before selecting any textile fibre for the intended use, the knowledge of the classification of textile fibres is essential.

Classification of textile fibres can be done in many ways. Some of them are as follows:

a. Classification according to their nature and origin.
b. Classification according to botanical/zoological/chemical name.
c. Classification according to the ability to attract water, i.e. moisture absorption.
d. Classification according to their thermoplasticity.
e. Classification according to their utility.

1.3.1 Classification according to their nature and origin

Charts 1.1–1.8 show the classification of main natural and manmade fibres used normally in textile applications. The fibres are normally classified as natural and manmade. Among natural subgrouping is made as animal origin, vegetable origin and mineral origin. The manmade fibres are further subgrouped as regenerated fibres and synthetic fibres.

Natural fibres are those fibres which are available from the natural sources, viz. plants, animals, minerals, etc. The mineral fibres are also referred as miscellaneous inorganic fibres.

Manmade fibres are those fibres which are developed by man. Man possesses a natural instinct of imitating nature and its products. Textiles are no exception to it. He does it either using some natural resources and/or chemicals to produce fibres, artificially. Therefore, sometimes back in their earlier development stages, manmade fibres were also called Artificial Fibres

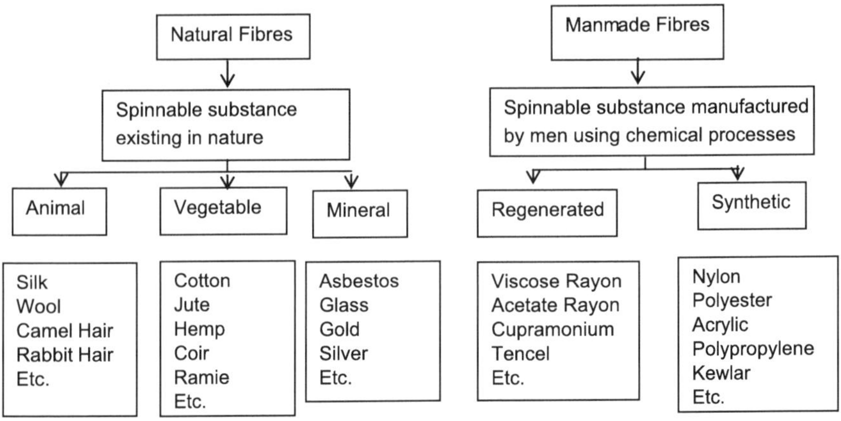

Chart 1.1 Classification of fibres.

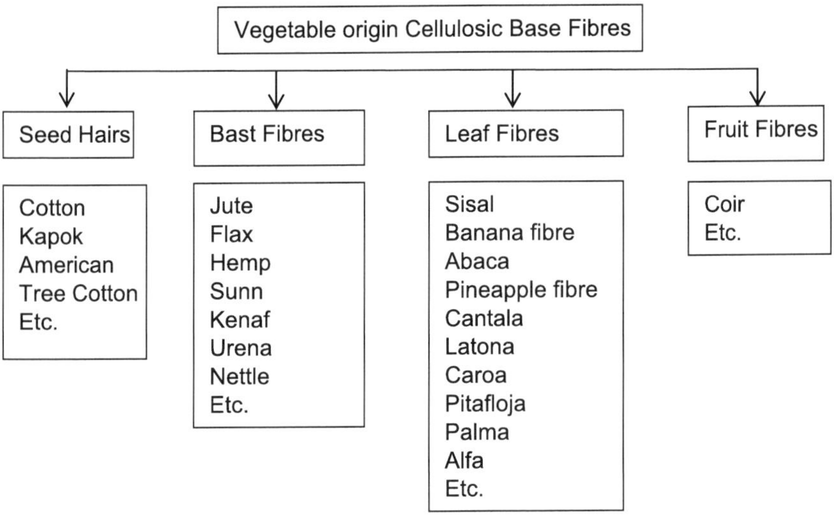

Chart 1.2 Natural vegetable fibres.

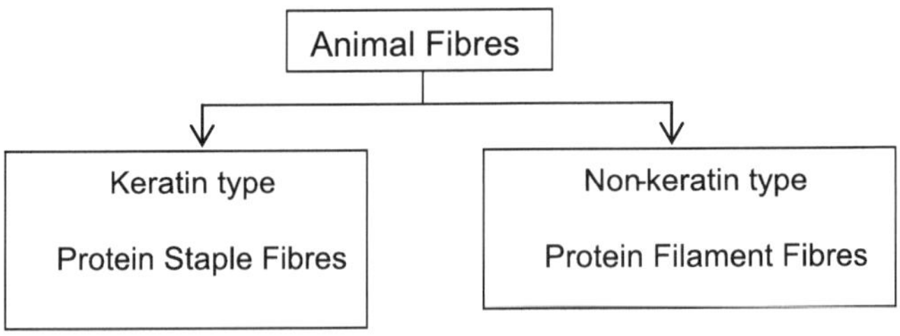

Chart 1.3 Animal fibres.

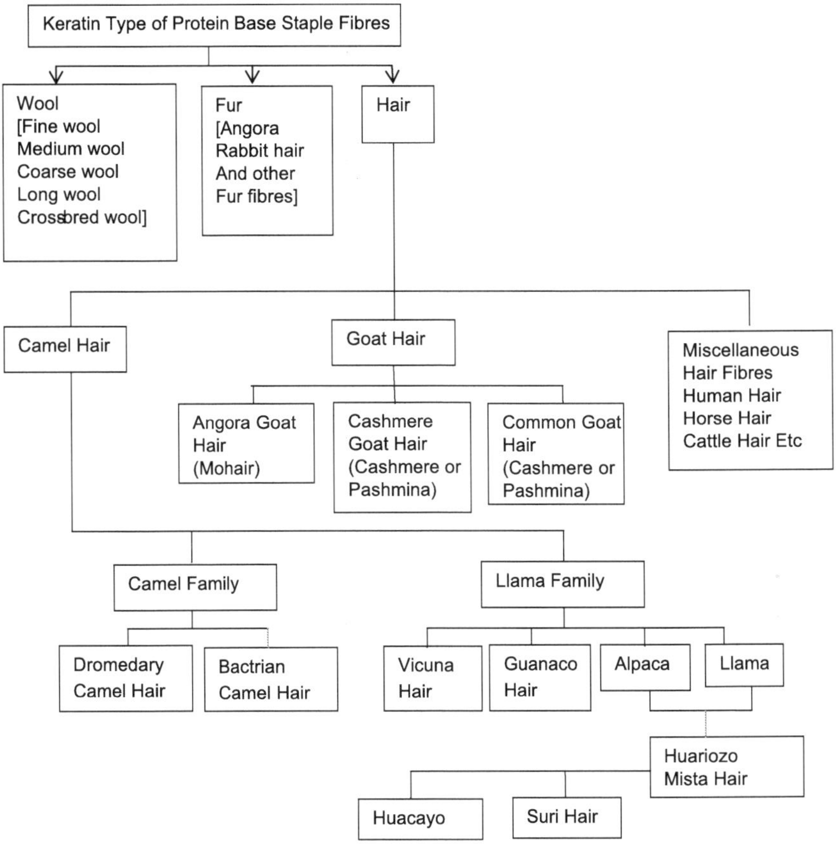

Chart 1.4 Keratin type fibres.

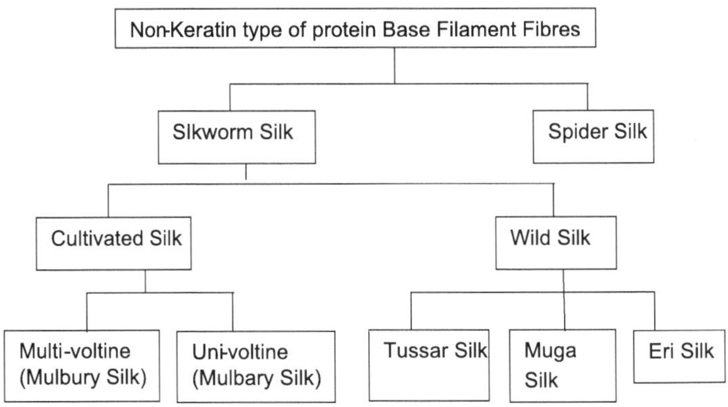

Chart 1.5 Non-Keratin type fibres.

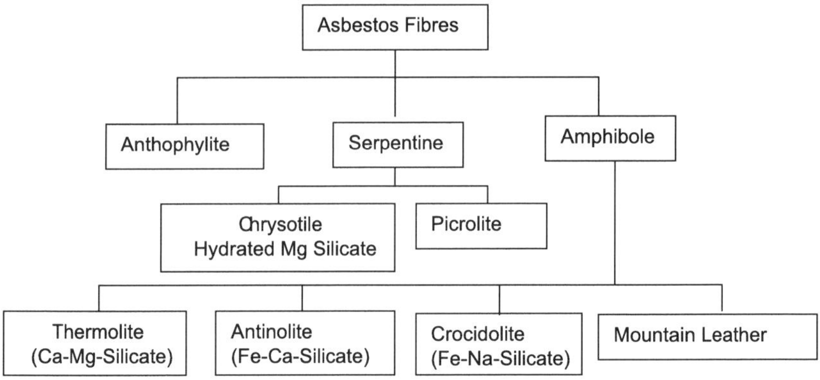

Chart 1.6 Mineral fibres asbestos fibres.

If the manmade fi bres are obtained from cellulosic base material, then they are termed as regenerated cellulosic manmade fibres; if they are manufactured from synthesizing using various chemicals like the petroleum products, then they are called synthetic manmade fi res.

Manmade fibres can be broadly classified into regenerated fibres, synthetic fibres and miscellaneous inorganic fibres.

Regenerated natural-polymer fibres (polymer is a fibre forming substance) are those fibres which are regenerated by using natural source as a base and are chemically shaped to filament form, e.g. viscose rayon, cuprammonium rayon, acetate rayon, casein ardil, etc.

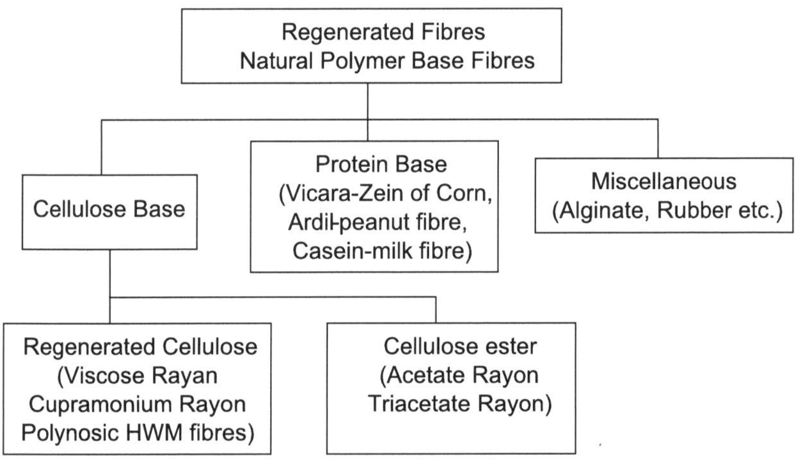

Chart 1.7 Manmade regenerated fibres.

Synthetic fibres are those fibres where only chemicals are used in the manufacture of such fibres. They do not require natural raw material as a base for the manufacture, as in the case of regenerated fibres. For example, hexamethylenediamine and adipic acid are used in the manufacture of nylon and dimethyl terephthalate and ethylene glycol in the production of polyester fibre etc.

The chemicals are converted into materials capable of forming fibres and these substances are manipulated into fibrous form.

In another way, synthetic fibres can also be classified in two groups, viz. (a) Heterochain fibres, e.g., polyester, polyamide, polyurethane, polyurea fibre, etc., (b) Carbochain fibres, e.g. polyacrylonitrile, polyvinyl alcohol, polyvinyl chloride, polyolefin and special purpose fluorine, etc., containing fibres.

The macromolecules of heterochain fibres contain in their main chain carbon atoms and atoms of other elements, such as, oxygen and nitrogen. These polymers are usually obtained by polycondensation or polymerization of cyclic compounds.

The macromolecules of heterochain fibres have a carbon skeletal chain, i.e., they contain only carbon atoms in the main chain. Such polymers are obtained by polymerization.

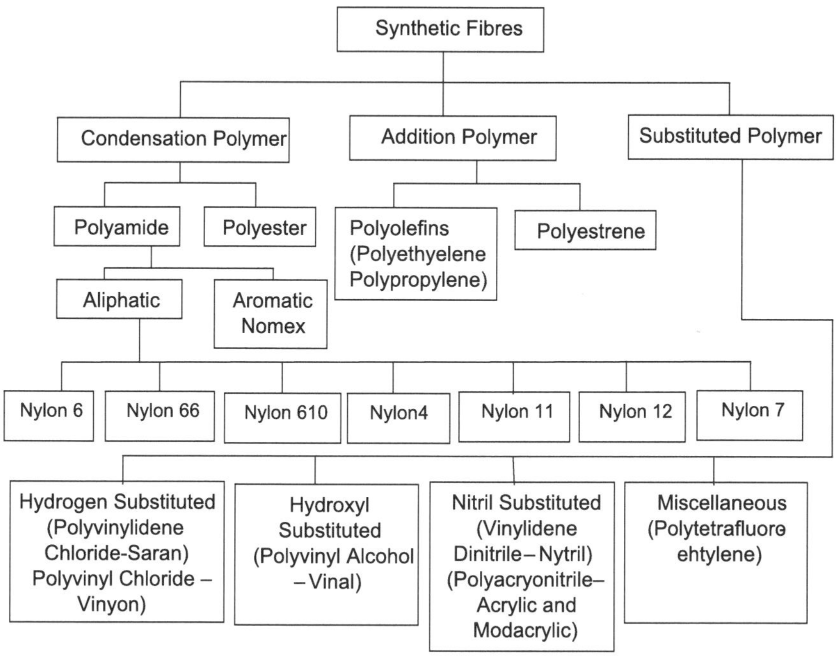

Chart 1.8 Manmade synthetic fibres.

Miscellaneous inorganic fibres are those fibres which are made from substances such as metal and glass. Both these materials have been used by man for a long time in forms other than textile fibres. Thus malleable and ductile nature suggested the use as textile fibres long ago, but the cost and technical difficulties hindered its wide use. Modern developments in converting both these materials into textiles have overcome the difficulties to ensure their modest regular use.

The classification of fibres under this category is by no means exhaustive but Charts 1.1–1.8 do include the main apparel use fibres as well as some of the less well known fibres.

1.3.2 Classification according to their botanical, zoological or chemical name

- Vegetable fibres are grouped under botanical classification. They include cotton, jute, flax, etc.
- Animal fibres are grouped under zoological classification. They include wool, silk and hair fibres.
- Chemical name is given to the classification of fibres under manmade fibres. The main constituent chemicals and mode of their production is explained in brief, for example, regenerated cellulose, polyamide linear macromolecules having in their chain the recurring amide functional group, etc.

Detailed classification is given in Table 1.2

Table 1.2 Varieties, geographical sources and uses of some important textile fibres

	Type	Varieties	Major sources	Use
		Natural fibres		
Vegetable origin		Botanical name		
1.	Cotton – Seed hair	Gossypium (G)		All grades of textile and cordage
	Upland type		USA, India and China	
	American type	*G. hirsutum*	West Indies and USA	
	Asiatic type	*G. herbaceum*	Egypt and Peru	
	Island type	*G. barbadense*	West Indies and Brazil, Haiti islands	
	Peruvian	*G. peruvianum*	Peru	
	American tree cotton	*G. purpurascend*	Argentina	

	Bast fibres			
2	Jute	*Corchorus olitorius* and *Corchorus capsularis*	India and Bangladesh	Carpets, rope, geotextile, hessian or burlap, sacks
3	Flax	*Linum usitatissimum*	USSR, Germany, France, Holland, America	Cordage and coarse textiles
4	Hemp	*Cannabis sotiva*	Philippines, North Bomea, etc.	Cordage and coarse textiles, clothing and nutritional products
5	Ramie	*Bohemeria nivea* and *Bohemeria tenacissima*	India, China, etc.	Cordage and coarse textiles, industrial sewing thread, packing materials, fishing nets and filter cloths
6	Abaca – Manila Hemp	Musa	Philippines, North Bomea, etc.	Cordage and coarse textiles
7	Sisal	*Agave sisalana*	Africa, Java, Mexico	Cordage and coarse textiles
8	Sunn	Fibre from the bast of *Crotalaria juncea*	India	Cordage and coarse textiles
9	Henequen	Fibre from the bast of *Agave fourcroydes*	Mexico	Cordage and coarse textiles
10	Maguey or *Agave americana or Century plant*	Fibre from the bast of *Agave cantala*	America, Asia,	Cordage and coarse textiles
11	Broom	*Cytisus scoparius* and/ or *Spartium junceum*	India, Europe	Mattress, bags
12	Kapok – Fibre obtained from the inside of the kapok fruit	Ceiba pentandra	Argentina	Fine textiles

13	Alfa Grass – Fibre obtained from the leaves	*Stipa tenacissima*	Northwest Africa and Spain	Floor coverings, home furnishing
14	Coir	*Cocos nucifera*	India	Cordages, home furnishing
Animal origin				
15	Wool or animal hair	Hair		
	Sheep	Ovisaries	World over	
	Goat hair (mohair)	Caprahircusan		
	Rabbit hair	Lepus	Europe	
	Alpaca	Aucheniapalo	Europe	
	Llama	*Auchenia llama*	Himalaya	
	Vicuña	*Auchenia vicuña*	South America	
	Camel Hair	Camelidae	Arabia, Sahara	
	Pashm or pashmina	Kashmir or Cashmere goat	Himalayas	Felt, textiles, cordage and coarse textiles
	Angora	Angora goat and Angora Rabbit	Europe	
	Yak	*Bos grunniens*	Himalayas	
	Guanaco	*Llama guanicoe*	South America	
	Beaver	Genus Castor	North America	
	Otter	*Lutra perspicillata*	India, Africa	
16	Silk			
	Cultivated silk	*Bombyx mori*	Japan, China, India	Textiles, apparels
	Wild silk	Eri, Muga, Tussah	India	
Manmade fibres				
	Regenerated fibres			
17	Viscose rayon	Regenerated cellulose fibre obtained by the viscose process for filament and discontinuous fibre	Produced under many trademarks in different countries	Textiles, light apparel, furnishing, interlining, carpets, etc.

18	Polynosic – Modal	Regenerated cellulose having a high breaking force and high wet modulus	Produced under many trademarks in different countries	
19	Cuprammonium rayon	Regenerated cellulose fibre obtained by the cuprammonium process	Bemberg	
20	Acetate rayon	Cellulose acetate fibre wherein less than 92% but at least 74% of hydroxyl groups are acetylated		
21	Triacetate	Cellulose acetate fibre wherein at least 92% of the hydroxyl groups are acetylated		
22	Alginate – Sea weed	Fibre obtained from metallic salts of alginic acid		
23	Protein	Fibre obtained from natural protein substances regenerated and stabilized through the action of chemical agents		
	Synthetic fibres			
24	Polyamide or nylon	Polyamide linear macromolecules having in their chain the recurring amide functional group	Manufactured in various countries	Technical textiles, floor coverings, sarees

25	Polyester	Polyester linear macromolecules having in their chain at least 85% (by mass) of an ester of a diol and terephthalic acid		Clothing and industrial purposes
	Acrylic	Fibre formed of linear macromolecules comprising at least 85% (by mass) in the chain of the acrylonitrilic pattern		Blankets, carpets and outerwear garments
	Modacrylic	Fibre formed of linear macromolecules having in their chain more than 50% and less than 85% (by mass) of the acrylonitrilic pattern		
	Chlorofibre	Fibre formed of linear macromolecules having in their chain more than 50% by mass of chlorinated vinyl or chlorinated vinylidene monomeric units		
	Fluorofibre	Fibre formed of linear macromolecules made from fluorocarbon aliphatic monomers		

	Polyethylene	Fibre formed of un-substituted aliphatic saturated hydrocarbon linear macromolecules		
	Polypropylene	Fibre formed of an aliphatic saturated hydrocarbon linear macromolecule where one carbon atom in two carriers a methyl side chain in an isotactic disposition and without further substitution		Ropes, insulation for electric cable, sanitary products. rugs and mats, under armour clothing, roofing membranes, filter media in water filters
	Polycarbamide	Fibre formed of linear macromolecules having in their chain the recurring ureylene (NH-CO-NH) functional group		
	Polyurethane	Fibre formed of linear macromolecules composed of chains with the recurring urethane functional group		

	Vinylal (m)	Fibre formed of linear macromolecules whose chain is constituted by poly(vinyl alcohol) with differing levels of acetilization		
	Trivinyl	Fibre formed of acrylonitrile terpolymer, a chlorinated vinyl monomer and a third vinyl monomer, none of which represents as much as 50% of the total mass		
	Elastodiene or spandex	Elastofibre composed of natural or synthetic polyisoprene, or composed of one or more dienes polymerized with or without one or more vinyl monomers, and which, when stretched to three times its original length and released, recovers rapidly and substantially to its initial length		Wide range of garments, especially in skin-tight garments.

	Elastane	Elastofibre composed of at least 85% (by mass) of a segmented polyurethane, and which, when stretched to three times its original length and released, recovers rapidly and substantially to its initial length		

1.3.3 Classification according to their ability to absorb moisture

From the point of view of wet processing the ability to bleach, mercerize, dye and give different finishes using chemicals to textile materials successfully depends on the ability to absorb moisture. The fibres which absorb moisture are called Hygroscopic or Hydrophilic fibres.

Hydrophilic fibres are characterized by the presence of hydrophilic groups which attract water. For example, all the natural fibres have groups in their molecules which attract water. Moisture absorption of hydrophilic fibres is higher than hydrophobic fibres.

Hydrophobic fibres are those which do not readily absorb moisture. All synthetic fibres, so far produced, contain very few water attracting groups. Absence of water attracting groups accounts for their low moisture absorption. The fibres which have lower moisture absorption are difficult to dye and bleach. Another disadvantage is that they develop static electricity charges quicker than hydrophilic fibres. This is an important factor which is responsible for some troubles during mechanical processing of fibres.

1.3.4 Classification according to their thermoplasticity

The textile fibres can also be classified into two types, viz.

a. Thermoplastic type

b. Non-thermoplastic type

Thermoplastic types are those which are deformable by heat and pressure, without accompanying chemical changes. This suggests that the thermoplastic types of fibres can be softened by heat, which means they can be moulded and heatset. The fibres which do not possess the above characteristic are designated as non-thermoplastic.

Most of the synthetic fibres have thermoplastic properties. Regenerated acetate rayon may melt when ironed hot, and polyvinyl chloride (PVC) fibres are most heat sensitive type. Some synthetic fibres have thermoplastic properties which are more pronounced than those of acetate, notably polyamide and polyester.

This property of thermoplasticity is used to heatset fabrics made from them and confer on them the dimensional stability. Also this quality is used to convert these fibres into new type of yarn such as Textured Yarn.

1.3.5 Classification according to their utility

The textile fibres can be broadly classified into two types under this category, viz.

1 Major textile fibres

2 Minor textile fibres

Major textile fibres are those which are widely used as textiles by the textile industry, e.g., cotton, wool, silk, jute, viscose rayon, acetate rayon, nylon, polyester, etc.

Minor textile fibres are those which are used to a very much less extent as textiles (by the textile industry), e.g. banana fibres, abaca fibres, asbestos fibres, bamboo fibres, soybean fibres, pineapple fibres, metallic fibres, milk fibres, casein fibres, alginate fibres, rubber, etc.

2
Some major natural fibres

Natural fibres are those that are available in nature. Cotton, wool, silk, jute are some of the major natural fibres which have attained a commercial importance in the fibre world.

Natural fibres are facing stiff competition in the fibre market following the introduction of manmade fibres.

Natural fibres are still holding their commercial status by virtue of some special characteristic properties which are not available with manmade fibres. Therefore, cotton fibre is still considered as 'King of fibres, the queen being the silk'.

Cotton is known for its cool wear; silk for its elegance; wool for its resilience and warmth; jute for its cheapness and durability in storing and transporting goods.

According to the historians India was the first country to grow cotton and weave cotton fabrics. The use of cotton for textiles was known to Indians from the time immemorial.

Wool was introduced in 80 AD in England. The art of manufacturing fine quality yarn and fabric was known to the people of Worstead – a village in England. Now the term worsted used in this trade is an adapted form of the word Worstead. Therefore, wool is the fibre, while worsted and woolen are the systems or the process of converting the wool fibre to the respective yarn, viz. woolen yarn (coarse), worsted yarn (fine).

Silk was known to China in 2640 BC. The Chinese monopolized the art for over 3000 years. But during the early period of Christian era the cultivation of the silk worm (or sericulture) gradually began to trickle out of China.

2.1 Advantages of natural fibres

The main advantage of natural fibres is that they are abundantly available in nature. Only the man has to cultivate the plants or rear the animals. They possess some unique fibre properties which make them superior to manmade fibres in certain respects. Some of the natural fibres are very cheap and easy to dye. They readily absorb moisture and perspiration, provide great bodily comfort, resist whole melting, generate no static charges during processing and produce no dermatological effect on human skin. It is easy to sew and the seams hold tight.

2.2 Disadvantages of natural fibres

Unlike manmade fibres, natural fibres are not available in high tenacity (HT) and medium tenacity (MT). Natural fibres do not possess high degree of resiliency as compared to manmade fibres, hence the fabrics made out of them do get wrinkles from 'baggy knees', possess less crease recovery.

Production of natural fibres cannot be completely controlled by man, in so far as the quality and quantity of the fibre is concerned. Therefore, from year to year quantities of specific qualities vary and this tends to cause price fluctuations according to variations in demand which cannot be foreseen until the fibres have been produced. The availability of natural fibres is affected by natural calamities and vagaries of nature.

The production of natural fibres involves the use of land which is also required for growing the agro products. With the availability of limited amount of land on the surface of the earth, the demand for land for food and housing on one hand and that for the growth of natural fibres on the other is to be balanced.

Variation in length, fineness, etc. of the natural fibre causes less regular and uniform yarn than that obtained from manmade fibres.

2.3 Cotton

Cotton provides about 50% of the world's textile fibre. The fibres are the outer covering of the seeds, which themselves yield valuable oil, used for the production of cooking oil or margarine or as a constituent of animal feeds.

The traditional varieties of cottons cultivated in this country (Indian subcontinent) in the past belonged to the species of *Gossypium arboreum* and *Gossypium herbaceum*, called generally as desi cottons. The varieties belonging to these species have inherently low yield and are shorter and coarser than varieties belonging to *Gossypium hirsutum* or American Upland types. Cottons belonging to *Gossypium barabadense* have extremely long and fine fibres and are grown mostly in Egypt and Sudan. In 1947–1948, the proportion of Indian cottons belonging to *G. hirsutum* was hardly 3% of the overall production, whereas today the cultivation of Hirsutum cottons has risen to above 50% of the overall area. The area under Barabadense, negligible even in 1970, has also started rising steadily. Apart from these, one more reason for the improvement in our cotton situation has been the introduction of a number of hybrid varieties of cotton which are known for their phenomenal increased yields compared to those of the older varieties.

The cotton plant grows in moderately hot climate and is widely distributed over the world. There are many varieties which produce fibres of different physical properties. The best qualities reaching the world market for superfine counts come from Egypt, however, the quantity grown are much less. A slightly shorter length quality is grown in India, China and Pakistan. It is of course possible to adapt the best qualities to various parts of the world so that cotton can be grown practically anywhere, provided the climate is suitable for the purpose.

Cotton is grown in 35 countries across the world and total production is around 900 million metric tons. The biggest cotton producers as on date are India, China, USA, Pakistan and Brazil. Table 2.1 gives the top 10 cotton producing countries for the year 2013–2014. In 2014–2015, India has emerged as the largest producer of cotton.

Table 2.1 Cotton Production in 2013–2014

Rank	**Country**	**Production in 1000 metric tons**
1	China	6967
2	India	6641
3	USA	2811
4	Pakistan	2068
5	Brazil	1633
6	Uzbekistan	904
7	Australia	893
8	Turkey	501
9	Turkmenistan	327
10	Greece	298
	Total World	**900,000**

Source: www.statista.com.

The qualities vary mainly in colour and staple length. American cotton is generally white as compared to Egyptian cotton which is light cream. Indian cotton is of a light grey colour. India is the only country growing all lengths of cotton starting from the lowest to the highest, but the majority of them have staple length usually not exceeding 25.4 mm, whereas Egyptian and American cottons grow up to 63.5 mm long. The lustre of the fibres also varies, Egyptian cotton being the most highly lustrous. It has also been found possible to grow coloured cotton (green, grey and brown) and such varieties are utilized in the USSR. It may be interesting to note that 90 mm length cotton fibre variety was obtained in Brazil by crossbreeding Sesido cotton with Peru-pima cotton [*Textile Research Journal*, 1980, 50 (12), 752].

The cotton plant belongs to the malaceae or mallow family. It is a perennial plant, with a taproot, cultivated since very ancient times (in the Nile valley by the Egyptians and in Central America by the Aztecs), for its textile fibres that come from the long bristles that encircle the seeds. It is perennial, that is, it produces seeds year after year. But after a time, the value of cotton deteriorates both in quality and quantity. The best results are obtained by cultivating it as an annual crop. This has also the advantage of enabling the cotton grower to improve his stock by the introduction of new strains.

Planting begins at various times in different regions, depending on the climate. In India it is from April to August.

The cotton seed is planted 1–10 cm deep in rows 1 m apart of the prepared land, depending on the moisture and the type of soil. If the soil is sufficiently warm and moist, young plants push up through the surface of the soil in 8–12 days. If it is too dry then germination may not take place and if it is wet and cold, then germination is slow and uncertain. The leaves and stems of the young plants are very tender, of fleshy colour and are smooth and oily. The leaves then become dark olive green and the plants grow rapidly. Within 40 days plant begins to form flower stalks. Flowering takes another 30 days. The opened flower is yellowish white on the first day when pollination occurs; it turns to pink on the next day when fertilization takes place and the petals of the flowers fall on the third day. The immature seeds thus formed, grow rapidly and the large cotton boll matures in 40–50 days.

Cotton fibres grow inside the closed pod. Pod contains cotton seeds, which are wrapped up in young, actively growing hairs. It is interesting to note that each cotton seed may produce as many as 20,000 fibres on its surface, and a single boll may contain 150,000 fibres or more. The fibres grow from the surface of the seeds from a single cell, which lengthens to a thin walled tubular structure of maximum length. The fibre wall is then thickened by daily growth of rings of cellulose.

Harvest occurs 6–7 months after seeding and the fibres are separated from the seeds (shelling). The fibres are assembled into balls for treatment in the textile industry. When the seeds are nearly ripe, the pod bursts open, and the cotton hairs project, forming a white fluffy mass which are usually accommodated in four sections in the pod. This is termed as boll. The fibres now complete their ripening, after which their cell contents gradually dry up, leaving the commercial cotton fibre. The drying of the cell contents under the influence of the sun is a very important factor in producing one of the chief properties of cotton, namely; its characteristic natural twist called convolution.

The number of convolutions varies according to the quality of cotton, e.g. Sea Island - 300, Egyptian - 230, American - 190, Indian - I50 per 25 mm.

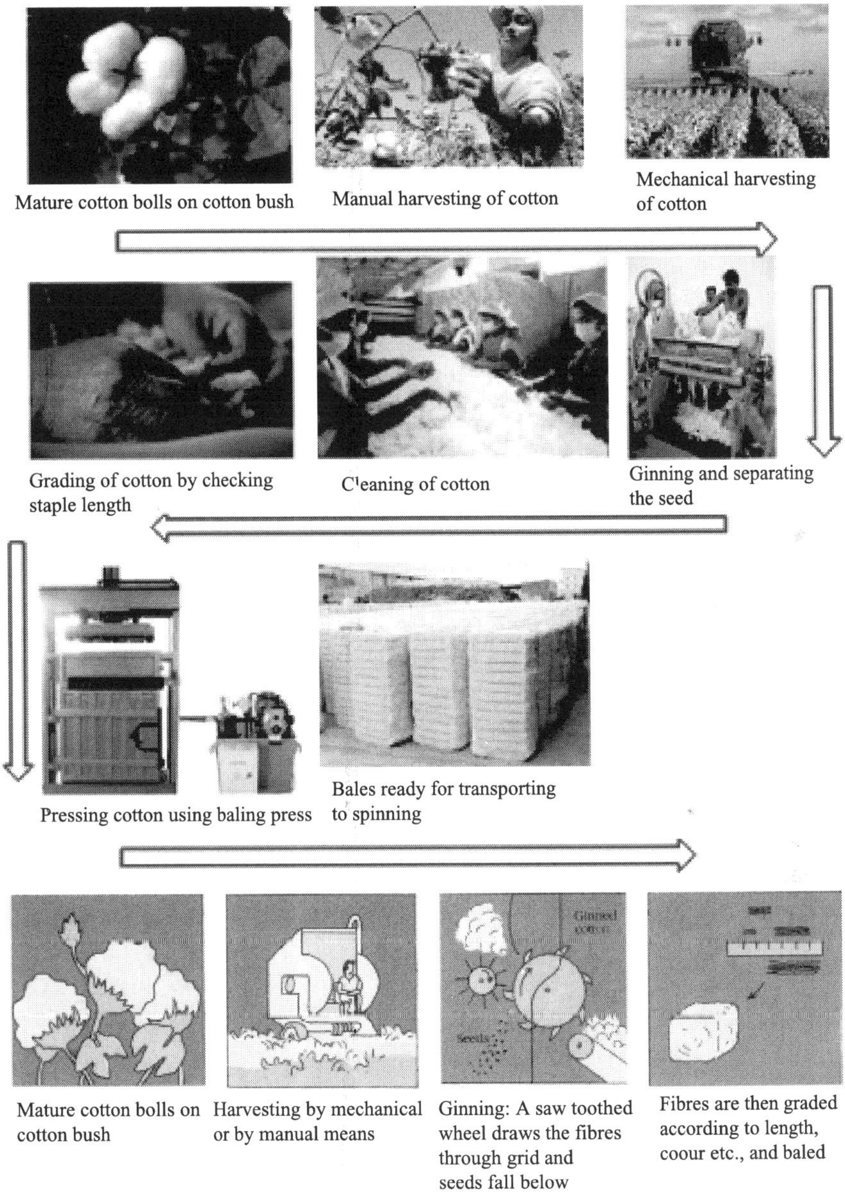

Figure 2.1 Flow chart showing cotton fibre production.

Cotton fibres are picked from the boll either by hand picking method or by machine picking method.

After the cotton has been gathered, the cotton fibres are separated from the seeds by a process known as ginning, in which it is passed between rotating knife roller or saw toothed discs, whereby cotton fibres are pulled off the seed. The seeds are then collected to be utilized for the manufacture of cotton seed oil, and the residue used as fodder. The fibres collected after the ginning process will have their upper free end tapering to a point, while basal end is open when it has been cut off from the seed. Fragments of the seed coat, leaves or stem, and accidental dirt may also be present in the ginned cotton. The ginned cotton is graded, depending on the quality of the fibre, before pressing them into bales. The fibres which are packed into bales are then dispatched to the spinning mills (see Fig. 2.1).

Fig. 2.2 gives the cotton production in different states as per statistics of 2012—

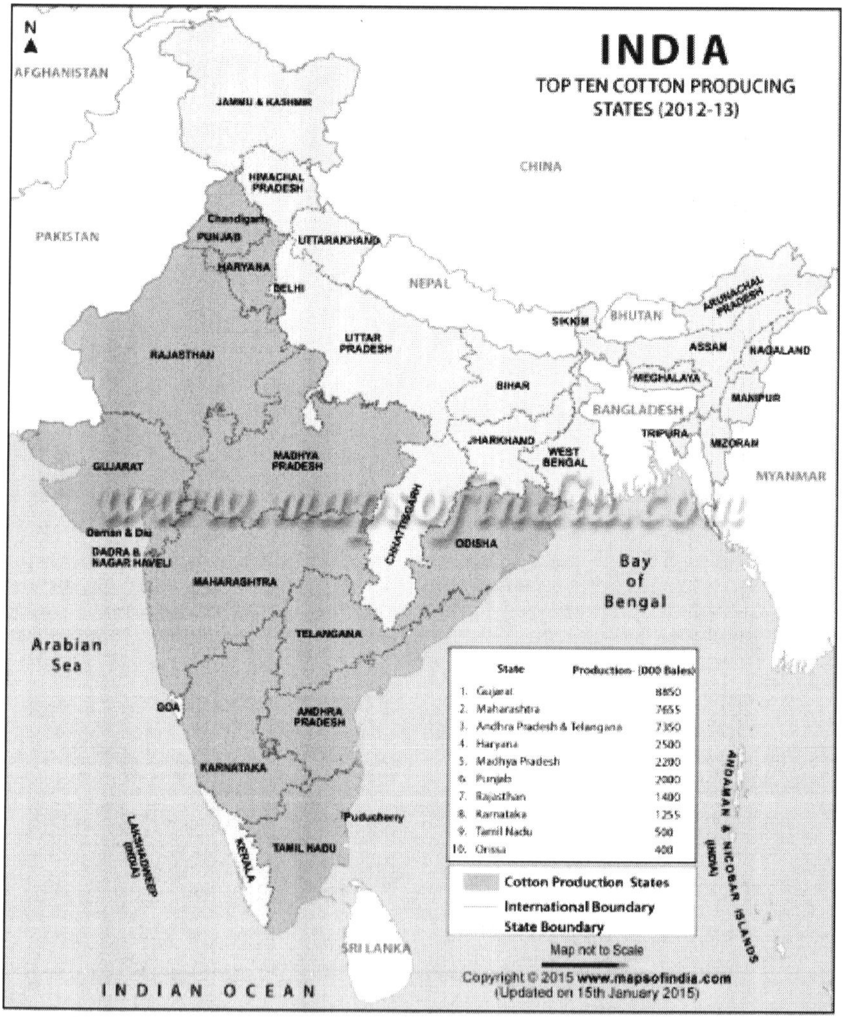

Figure 2.2 Top cotton producing states.

In India, the states of Maharashtra, Gujarat and Andhra Pradesh are the leading cotton producing states. These states have a predominantly tropical wet and dry climate.

2.3.1 Bt-Cotton

Bt-Cotton pests attack these crops causing extensive damage. Pest attack accounts for the major cause of reduced production. There are about 162 species of insects, which are known to devour cotton at various stages of growth, of which 15 are considered to be key pests. Among the insects, cotton bollworms are the most serious pests in India. Over the past 40 years, many pests have developed resistance to pesticides.

The only successful approach to engineering crops for insect tolerance has been the addition of Bt-Cotton toxin, a family of toxins originally derived from soil bacteria. Thus genetically engineered form of natural cotton is produced by inserting a synthetic version of a gene from the naturally occurring soil bacterium Bacillus *thuringiensis*, into cotton. The primary purpose is to induce the plant to produce its own BT toxin to destroy the bollworm, a major cotton pest. The gene causes the production of BT toxin in all parts of the cotton plant throughout its entire lifespan. When the bollworm ingests any part of the plant, the BT cotton toxin pierces its small intestine and kills the insect. The BT toxin contained by the BT crops is no different from other chemical pesticides, but causes much less damage to the environment. These toxins are effective against a variety of economically important crop pest but pose no hazards to non-target organisms like mammals and fish. Three BT crops are now commercially available: corn, cotton and potato. The main selling points of Bt-Cotton are the reductions in pesticides to be sprayed on a crop and the ecological benefits which stem from that.

The efforts to reduce the loss in yield got diverted to harness genetic engineering technology for bollworm resistance in India in 1990s. The main advantage of utilizing biotechnology in agriculture are the possibilities of increase in productivity through the use of newer varieties that posses properties such as resistance to pests, diseases and other stressful conditions like drought, salinity or water logging. The chronology of Bt-Cotton development started with the formation of Institutional Biosafety Committee (IBSC) and application for transgenic Bt-Cotton seed import in 1994. As of now, cotton is the most popular of the BT crops: it was planted on about 1.8 million acres (728,437 ha) in 1996 and 1997. The BT gene was isolated and transferred from a bacterium Bacillus thurigiensis to American cotton. The American cotton was subsequently crossed with Indian cotton to introduce the gene into native varieties. In April 2002, GEAC (Genetic Engineering Approval Committee) approved three Bt-Cotton hybrids for commercial cultivation in six states (Andhra Pradesh, Gujarat, Karnataka, Madhya Pradesh, Maharashtra, Tamil Nadu) after taking into account the data on their performance.

Field trials have shown that farmers who grew the BT variety obtained 25–75% more cotton than those who grew the normal variety. Also, Bt-Cotton requires only two sprays of chemical pesticide against eight sprays for normal variety. As of now about 62 Bt-Cotton hybrids have been developed by private seed companies, which are under commercial cultivation after the approval of GEAC.

The use of Bt-Cotton in India has grown exponentially since its introduction. Recently India has become the number one global exporter of cotton and the second largest cotton producer in the world. India has also bred Bt-Cotton varieties such as *Bikaneri Nerma* and hybrids such as (NHH-44), setting up India to benefit now and well into the future. Socio-economic surveys confirm that Bt-Cotton continues to deliver significant and multiple agronomic, economic, environmental and welfare benefits to Indian farmers and society including halved insecticide requirements and a doubling of yields. Bt-Cotton now accounts for over 94.75% of the country's entire cotton cultivation (Ref. "The Hindu" – 24 September 2013).

2.3.2 Organic cotton

Research work at the Central Institute for Cotton Research was translated into an organic cotton production technology, focusing on 'on-farm' recycled soil amendments and biorational pest management. Organic cotton is also known as green cotton, environmentally friendly cotton or biological cotton.

As per the National Organic Standards Board, USA, organic agriculture is defined as an ecological production management system that promotes and enhances biodiversity, biological cycles and soil biological activity. It is based on minimal use of 'off-farm' inputs and no management practices that restore maintain and enhance ecological harmony.

Organic cotton is grown and processed without toxic chemicals like pesticides, fungicides, plant growth regulators and synthetic fertilizers that can be absorbed easily when in contact with the user's skin. Pesticides, fertilizers and chemicals used to grow and process conventional cotton fabrics may go directly to the user's blood stream, which consequently affects the body organs and tissues. Table 2.2 gives the advantages of growing cotton organically.

Table 2.2 Advantages of cultivating cotton organically

	Conventional cotton	Organic cotton
Environment	• Pesticides kill beneficial insects • Pollution of soil and water • Resistance of pests	• Increased bio-diversity • Eco-balance between pests and beneficial insects • No pollution
Health	• Accident with pesticides • Chronic diseases (cancer, infertility, weakness)	• No health risks from pesticides • Healthy organic food crops
Soil fertility	• Risk of declining soil fertility due to use of chemical fertilizer and poor crop rotation	• Soil fertility is maintained or improved by organic manures and crop rotation
Market	• Open market with no loyalty of the buyer to the farmer • Dependency on general market rates • Usually individual farmers	• Closer relationship with the market partner • Option to sell products as "organic" at higher price • Farmers usually organized in groups
Economic	• High production costs • High financial risk • High yields only in good years	• Lower costs for inputs • Lower financial risk • Satisfying yield once soil fertility has improved
(Courtesy: Globe Cotyarn Pvt. Ltd.)		

Organic production systems replenish and maintain soil fertility, reduce the use of toxic and persistent pesticides and fertilizers and build biologically diverse agriculture. To increase organic matter in the farm crop rotation/crops used for mulching, green manuring, manure/compost should be used in balanced proportion. Intercropping is used where secondary crops (often sunflowers or millet) are grown between and around small plots of cotton. To control weeds, use of green manuring, crop mulching and hand weeding is recommended.

To control pest and diseases use of resistant varieties, follow timely sowing, intercultural operations, cultivation of intercrop and trap crops, use of pheromone traps, use of parasites and predators, use of bio-control that is *Bacillus thuringiensis*, use of herbal oils and recommended pesticides (pyrethrum, retanon) is done. There are rules for organic cotton processing and preparing fabrics.

Organic cotton is mostly a consumer driven movement. Three organic cotton projects are in operation in various states like Gujarat, Maharashtra and Madhya Pradesh. For organic **"farming certification"** is important. Agencies giving certification should ensure that in the land which is to be used for organic cotton farming, from last three years there is cultivation without pesticides and chemical products or artificial fertilizers. If there is used of chemicals and pesticides at the moment than one has to wait for a period of three years before the chemical residues on the farm disappears. Organic cotton is not a variety; it is the more of certification of the non-toxicity in soil of the agricultural land and the produce from it.

Fig. 2.3 gives a clue for successful organic cotton farming as explained by Globe Cotyarn Pvt. Ltd.

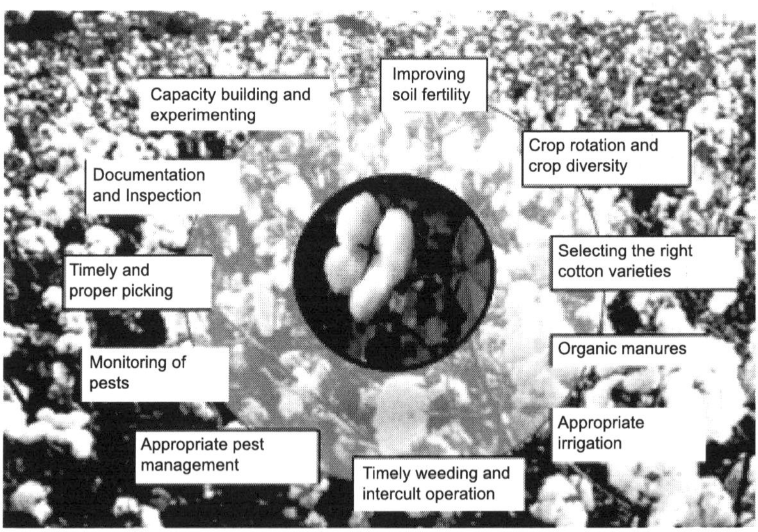

Figure 2.3 Successful organic cotton farming with a system approach.

The ingredients for success need to be applied together are:

1. Suitable measures to improve and maintain soil fertility;
2. Establishment of crop rotation and crop diversity, fostering natural balance;
3. Selection of varieties suitable to the conditions (soil, availability of irrigation, market requirements);
4. Appropriate types and amounts of manures at the right time;
5. Timely crop management such as intercultural operations, weeding and irrigation;
6. Careful monitoring of the crop and sufficient protection against pests according to the concept of economic threshold level;
7. Timely and proper picking of the cotton;
8. Sufficient documentation for inspection and certification;
9. Capacity building and experimenting for continuous improvement.

In 2000–2001, international production was approximately 6368 metric tons (slightly more than 14 million pounds, or 29,248 bales), grown in 12 countries, according to data from the Pesticides Action Network of the United Kingdom and from the Organic Trade Association (OTA). In India, net income per acre on Bt-Cotton was lower by 37% at Rs. 3351 compared to organic cotton, which stood at Rs. 5618 thus as compared to Bt-Cotton, organic cotton yield is more. Organic cotton still only occupies a tiny niche of far less than 1% of global cotton production. However, the number of farms converting to organic cotton and the number of projects is constantly increasing. Besides the naturally soft organic cotton fabric is a lot more comfortable to use and is used in organic mattresses and is available at competitive prices. Though organic cotton cultivation and farming is found difficult but it can be adoptable by following suitable and careful cultivation techniques. For this, research, appropriate guidance and efforts and willingness of farmers are must. Also there is need of infrastructure for cotton purity and marketing.

The global organic cotton fibre supply has increased to 392% since 2000–2001 harvest to 25,394 metric tons during the 2004–2005 crop years. Supplies are projected to grow to 31,017 metric tons (68,237,400 pounds or 142,161 bales) by the end of the 2005–2006 harvest, reflecting an annual growth rate of 22%.

During the 2004–2005 harvest, cotton was produced in 22 countries with Turkey growing 40%, India 25%, the United States 7.7% and China 7.3%, respectively. In 2005–2006, these four countries combined are projected to produce 79% of the global organic cotton fibre from the projected 2005–2006 crop is sold or otherwise committed.

All existing organic cotton producer groups are expected to maintain or slightly expand production for the 2006–2007 harvest, and a small number of new projects growing cotton for the general market are expected to begin production in 2006–2007. Additional projects which are growing cotton for specific supply chains are expected to come on line beginning in the 2007–2008 harvest.

Global retail sales of organic cotton apparel and home textile products reached an estimated $3.2 billion in 2008 (Source: Organic Exchange Market Report 2007–2008). This represents a 63% increase from $1.9 billion market in 2007. The top ten organic cotton using brands and retailers globally were Wal-Mart (USA), C&A (Belgium), Nike (USA), H&M (SE), Zara (Spain), Anvil (USA), Coop (Switzerland), Pottery Barn (USA), Green Board (USA) and Hess Natur (Germany).

Most of the brands and retailers selling organic cotton products remain committed to their sustainability plans to expand their product lines by 24% in 2009 and 33% in 2010 with an estimated $4 billion and $5.3 billion market, respectively.

According to organic exchange organic cotton farms and fibre reports 2008, organic cotton production increased to 145,872 metric tons grown on 161,000 ha in 22 countries worldwide produced in 2006–2007.

It may be noted that the organic cotton production is based on a system of farming that maintains and replenishes soil fertility without the use of toxic and persistent pesticides and fertilizers or genetically modified seeds.

2.3.3 Coloured cotton

There is worldwide concern about the growing health hazard associated with the use of textiles dyed with the azo-group and other synthetic dyes.

A new trend has emerged in the cotton market scenario as the natural coloured cotton has caught the fancy of crop scientists. Natural colour cotton is a naturally pigmented fibre that grows in shades of green, brown and beige. The cotton plants are also hardy and naturally resistant to pests, eliminating the need for environmentally damaging pesticides and insecticides. The whole process from planting to growing to spinning to readymade garments uses no dyes, toxic or harmful materials. It is 100% pure and natural. It is the most environmentally green product possible and creates a beautiful, breathable, pure fabric with a soft, delicate feel. The natural colour is due to the plant's inherent (non-modified) genetic properties. University of California and Bureau Veritas testing have shown that it compares favourably to white cotton fabrics in a study evaluating abrasion, resistance, dimensional stability and pilling resistance.

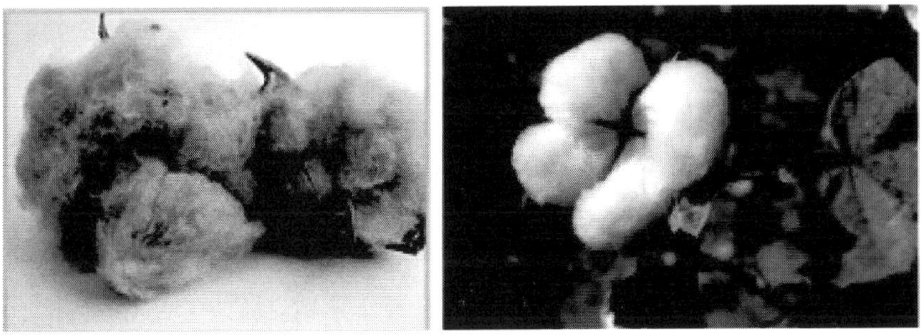

Coloured cotton Normal cotton boll

Figure 2.4 Naturally coloured cotton.

Colour grown cotton naturally grows in shades of creams, greens and browns so no further dye is required. The renewed interest in research on colour cotton is an off-shoot of the growing consumer driven movement as it is an eco-friendly alternative to dye-based textiles. Earlier the interest was slim in this due to low economic spin-off and inferior quality of cotton, but researchers now have come up with commercial varieties by incorporating desirable strains through breeding techniques.

The natural coloured cotton is much more favourable and desirable as compared to the chemically dyed cotton as in the later case colour fades when treated with detergents. Recently an indigenous cotton plant "*Gossypium arboreum*" is identified that possessed medium stable length and good strength, which earlier was the major drawback of coloured cotton. Various shades of brown and green have also been identified in "*Gossypium hirsutum*" and "*Gossypium barbadense*" variety of cotton.

Since naturally coloured cotton does not have pesticides, chemicals, viz. bleaches, dyes, fewer allergies and respiratory problems are found. Naturally coloured cotton feels softer to the skin and pleasant smell. Naturally coloured cotton is still relatively rare commodity because it requires specialized harvest techniques and facilities, making it more expensive to harvest than normal white cotton.

Yields of naturally coloured cotton are typically lower and the fibre is shorter and weaker but has softer feel than the more commonly available 'white' cotton.

2.4 Jute

Jute is a major commercial natural fibre of cellulosic origin, vegetable type bast fibre. Bast fibres form bundles or strands that act as ropes in the fibrous layer beneath the bark of *Dicotyladenous* plants. A number of small fibre cells are joined together, to form a long commercial fibres, by non-cellulosic materials usually lignin (see Fig. 2.5).

Jute is the second most widely used fibre, exceeded only by cotton. It is one of the oldest fibres used by the man.

Colour of jute fibre varies from yellow to brown to dirty grey. It possesses natural silky lustre. Generally jute fibre is coarse and is harsh to feel, though best qualities are soft and smooth. Jute is more durable than flax and cotton especially in respect of liability to deterioration by bacterial attack under damp conditions. Jute fibre is not as strong as flax or hemp. Individual fibres vary greatly in strength owing to the irregularities in thickness of cell walls.

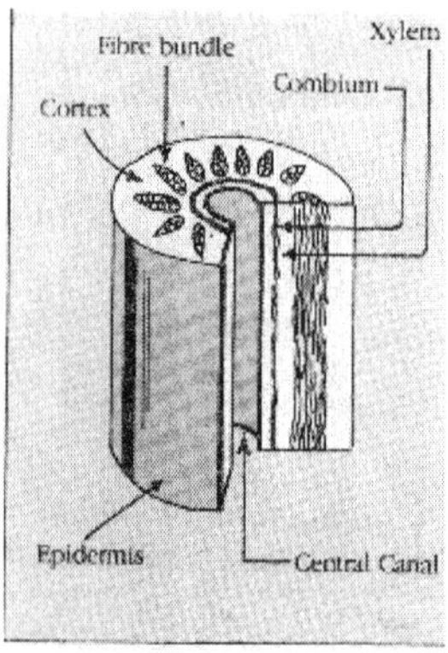

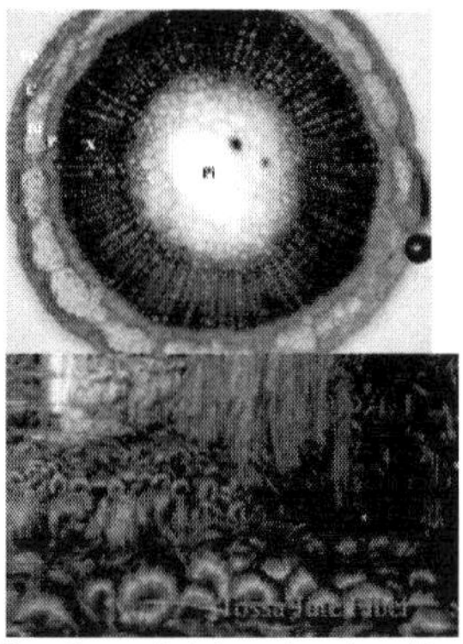

Bundles of jute, showing the fibres of *Corchorus olitorius* (tossa jute fibre) and *Corchorus capsularis* (white jute fibre)

Jute plant

Figure 2.5 Anatomy of the jute stem.

Jute is an annual plant growing from 5 to 10 ft high. It has a cylindrical stalk as thick as a man's finger. There are no branches except near the top. The crop is ready for cutting when the flowers begin to fade. The fibres are formed of bundles of cells cemented together and the individual cells are very similar in length and fineness to cotton fibres. The cheapest method of removing the fibres from jute plant is to steep it in the streams or ponds. This process is known as retting. After retting, the bundles are dried and then softening and scutching are done. Scutching is a mechanical beating or breaking process whereby fibres are separated intact from the long bundles of fibres.

Jute is cheap and is reasonably strong. This makes jute an important fibre for sacks and packing cloths. Jute is largely used for transporting and storage of products. This use of fibre has been possible and practicable due to jute's high resistance to stretching. Most familiar uses of jute are clothing fibres which are mainly of jute mixed with wool. Finer qualities of jute fibres are used in furnishing cloth. Jute is also used in cattle beddings, backing for tufted carpets and carpets and lanoleums, tarpaulins, ropes, foot-rugs, etc. Bituminized jute fibres, called prefabricated bituminous surfacing (PBS), may be used for rapid surfacing of runways, roads, roofing felts, etc. Jutelac may be used as substitute for plywood and black sheet. Main jute producers in 2012 are as given in Table 2.3.

Table 2.3 The top 5 jute producing countries

	Country	Jute production (2012)	% of world total
1	India	1,910,000 m/t	55.1%
2	Bangladesh	1,452,044 m/t	41.9%
3	China	45,500 m/t	1.3%
4	Uzbekistan	20,000 m/t	0.57%
5	Nepal	14,424 m/t	0.41%

Source: www.top5ofanything.com.

2.5 Wool

Wool is the fibre derived from the specialized skin cells, called follicles, of animals in the Caprinae family, principally sheep, but the hair of certain species of other mammals such as goats, llamas and rabbits may also be called for wool. Wool has several qualities that distinguish it from hair or fur: it is crimped; it has a different texture or handle; it is elastic; and it grows in staples (clusters).

Wool's scaling and crimp make it easier to spin the fleece. They help the individual fibres attach to each other so that they stay together. Because of the crimp, wool fabrics have a greater bulk than other textiles and retain air, which causes the product to retain heat.

Wool is a keratinous type protein base staple fibre. Keratin fibres are subdivided into three categories, viz.

1. Wool fibres which are grown by sheep.
2. Hair fibres which are grown by animals other than sheep.
3. Fur fibres which are grown by rabbit, mink, etc.

Wool comprises about 33% keratin, 26% dirt, 26% suint (usually salts of sodium, potassium or fatty acids), 10% wool fats and 5% mineral impurities. Keratin consists of more than 20 different types of amino acids.

The amount of crimp corresponds to the thickness of the wool fibres. A fine wool like Merino may have up to a 100 crimps/in., while the coarser wools like karakul may have as few as 1–2 crimps/in.. Hair, by contrast, has little if any scale and no crimp, and little ability to bind into yarn. On sheep, the hair part of the fleece is called kemp. The relative amounts of kemp to wool vary from breed to breed, and make some fleeces more desirable for spinning, felting or carding into batts for quilts or other insulating products.

Wool is the protective gown of sheep Fig. 2.6. Sheep live in many parts of the world and varieties can be adapted to practically any climate. Australia produces nearly 1/3 of wool produced in the world, stands first in the production and export of raw wool, and there are as many as 152 million sheep which is about 12 times the total human population. The average yield of raw wool per year per sheep in most of the countries is about 2.5 kg. In spite of having about 4.3% of the sheep population, India produces only about 1.2% of wool produced in the world. This is due to low yield of wool by the Indian sheep.

One fleece produces various qualities of wool fibre. They are numbered 1 to 14 as shown on the diagram indicating the length, softness and fineness

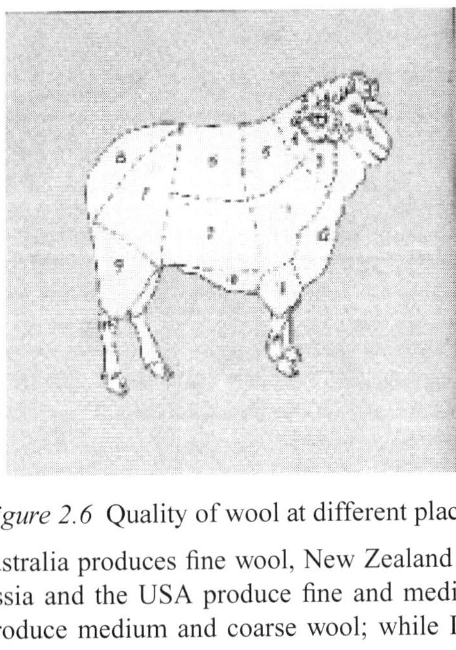

Figure 2.6 Quality of wool at different places.

In general, Australia produces fine wool, New Zealand produces fine and coarse wool, Russia and the USA produce fine and medium wool, the UK and Argentina produce medium and coarse wool; while India and Pakistan produce coarse wool.

Each variety of sheep produces a different type of wool; the Merino sheep, which originated in Spain, but which is now reared in many parts of the world, produces one of the best qualities of wool fibre.

Depending upon the breed of sheep, the quality of wool fibre varies. One distinguishes fine wool, medium wool and coarse fibres roughly corresponding to the designations Merino wool, crossbreed wool and carpet wool. Broadly we can classify wool into three categories according to their quality, namely, (1) Fine wool, (2) Medium wool and (3) Coarse wool.

Wool possesses much greater ability to return to its natural length after being stretched than any artificial fibre. In water or steam, wool can stretch to about twice its length without breaking. Because of elasticity, woolen garments may be stretched or crushed without losing their shape. Also, this fibre felts readily by heat, moisture and pressure being applied during the manufacturing processes.

Wool fibres are hygroscopic. This means wool has the power to readily absorb and give off moisture. It can absorb moisture up to one-third of its own weight. Synthetic fibres can only absorb 2–3% of moisture before they begin to feel wet. Additionally, wool absorbs sound and reduces noise level considerably like many other fabrics. Wool is generally a creamy white colour, although some breeds of sheep produce natural colours such as black, brown, silver and random mixes.

Wool is harder to ignite than most synthetic and cotton fibres used in equivalent products (higher ignition temperature); it has lower rate of flame spread, low heat release and low heat of combustion; does not melt or drip; forms a char which is insulating and self-extinguishes and contributes less to toxic gases and smoke than other flooring products when used in carpets. Wool carpets are specified for high safety environments such as trains and aircraft. Wool is usually specified for garments for fire-fighters, soldiers and others in occupations where they are exposed to the likelihood of fire.

Wool is static resistant as the retention of moisture within the fabric prevents a buildup of static electricity. Woolen garments are much less likely to spark or cling to the body. The use of woolen car seat covers or carpets reduces the risk of a shock when a person touches a grounded object.

2.5.1 Wool quality number

Wool quality number is nominally intended to indicate the highest worsted count to which it can be spun. It is probably more accurately interpreted as implying a certain fineness of fibre. This has arisen owing to the importance of fineness as a factor affecting the spinning quality, and to the consequent stress which is laid upon it by the wool classers.

From the foregoing discussion, and from the fact already mentioned that a single fleece varies considerably within, it will be realized that at least, so far as raw wool is concerned, quality numbers should be taken as being only general indications of probable spinning performance.

In the trade, a series of numbers called quality numbers is, therefore, used to indicate the difference in standards. The highest number is given to the finest wools, the lowest to the coarsest wools. The normal range of commercial wool is from 80^s quality down to 28^s quality, the numbers being apportioned to conform to easily distinguishable differences in standard, particularly fineness. The numbers are not regularly progressive or spaced but the full range as 90, 80, 70, 64, 60, 58, 56, 54, 52, 50, 48, 46, 44, 40, 36, 32 and 28^s. These numbers should not be confused with those used for the gauge or thickness of a yarn. In yarn processing, such numbers are known as yarn count or spinning count. This yarn number is one which can be absolutely defined and determined, whereas quality fineness number of fibre cannot be defined and determined. But quality fineness number cannot so accurately be related to the mathematical measurement of the diameter of the fibres, e.g. from 64^s quality maximum of 48^s and minimum 30^s yarn count can be spun.

2.5.1.1 Fine wool

Merino sheep, generally Australian merino, Rambouillet, American merino breeds produce fine wool. The quality (designated as wool quality number) varies from 60^s upward length of fibre varies from 4 cm to 12 cm for one year growth. These fibres are non-medullated and the diameter of fibre is below 24 μm.

2.5.1.2 Medium wool

Some breeds of sheep such as Hampshire, Southdown, Cheviot, produce medium quality wool. Wool quality varies from 48^s to 60^s. Length of fibre varies from 5 to 10 cm for one year growth. Diameter varies from 24 to 32.5 μm. These fibres are slightly medullated.

2.5.1.3 Coarse wool

Most of the Asiatic breeds of sheep produce coarse wools which are invariably medullated. They are mostly of quality below 48^s, and their length varies from 6 cm to 15 cm for one year growth. Diameter ranges from 32.5 μm and above. Since these wools are used mainly in carpets, they are often known as carpet wools.

In addition to the above mentioned varieties of wool there are two other important varieties namely, long wool and cross-breed wool.

2.5.1.4 Long wool

Breeds such as Cotswool, Romneymarsh, Lincoln, etc., which are generally regarded as mutton type of sheep, which produce medium to coarse wools which are very long, i.e. about 12 cm–40 cm for one year growth. These wools may be medullated and their quality usually will be in the range of 36^s–50^s and their diameter ranges from 30 to 40 μm.

2.5.1.5 Cross-breed wool

These are not pure breeds. These breeds are produced from two different breeds of sheep. Depending upon the parent breeds of sheep, the characteristics of the cross-breed will vary.

Corriedale and Polwarth are typical examples of cross-bred sheep. Their quality varies from 48^s to 64^s. Their length varies from 6 cm to 17 cm for one year growth, and the diameter ranges from 22 to 32.5μm.

The wool fibres are obtained by shearing or clipping the sheep in such a manner that the coat of hair or fleece remains in one piece. Shearing may take place once or twice a year.

The number of fibres in the fleece varies from about 1 billion in the fine merino to some 15–20 millions in the coarse wool breeds.

The number of fibres grown in a given area of skin is not uniform over the whole of the skin. The density is greatest around the line of spine and is decreasing towards the belly. The average weight of fleece is approximately 4.53 kg, but it may be as low as 1.36 kg or as high as 11.34 kg.

The low quality is also determined by the part of the body from which the wool comes, that of the sides, stomach, and of other larger spaces being more valuable than that of the smaller areas like back of the neck, legs, tail, etc.

Factors influencing the quality are the staple length, crimp, lustre and uniformity. All these properties depend on the climatic conditions, quality of soil, pasture, cultivation, shearing, etc.

The wool of dead sheep is of inferior quality, especially if it is removed from the skin with the aid of lime water. This procedure softens the roots sufficiently to allow easy removal of the hair, and such wool is called pulled wool

Most of the wool-clip contains added, acquired and natural impurities. The wool clips are packed into bales, which are then dispatched to the spinning mills.

Though wool is obtained from the sheep in many countries it is more difficult and more expensive to produce than cotton so that the total amount of wool available is considerably lower than that of cotton, therefore the basic price is approximately three times that of cotton (on average) because demand is always high.

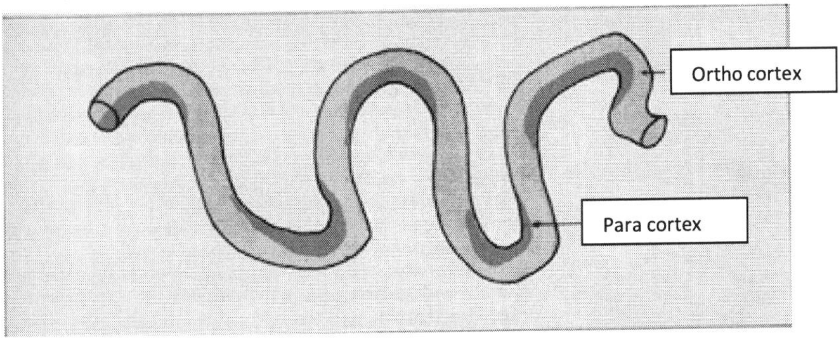

Figure 2.7 A wool fibre showing its spiraling crimp. The para-cortex always tends to be on the inside of the spiral.

It is well known that wool is a much longer fibre than the cotton fibre. The wool fibre possesses a feature called crimp which is a permanent wave (Fig. 2.7). Fine wools are very crimpy; the crimp becomes less evident as the wool becomes coarser. Wool fibres have to be subjected to more complicated series of processes than cotton fibres to convert them into yarns and fabrics.

Figure 2.8 Worsted and woollen yarn.

a. Smooth worsted yarn from long staple fibres.

b. Uneven woollen yarn from short staple fibres.

Wool fibre can be processed both on woollen system and on the worsted system. Worsted is, therefore, not a fibre but a system of spinning wool fibres (see Fig. 2.8). Wool can be seen as a fibre with outstanding clothing properties of softness, warmth, resilience and drape.

The warmth of fine wool fabrics is due to thousands of tiny air pockets trapped in the yarns because, the crimpy resilient fibres stand away from each other, and the fibre itself being protein does not transmit heat quickly.

A wool fabric acts as an insulating medium preventing a quick escape of body heat so that a feeling of warmth is engendered. A further aid to comfort is that wool can absorb excess moisture without feeling cold and clammy as cotton does in similar circumstances.

2.6 Silk

Silk is usually referred to as the queen of the fibres. It is a product of the life cycle of the silk cocoon. There are two major varieties of silk, viz. (a) cultivated silk and (b) wild silk. In the case of cultivated silk, the life cycle is well controlled, starting with the eggs which are incubated in specially tempered rooms, and which have to be chosen according to quality, right up to the selection of the spinner and cultivation of the mulberry plant nothing is left to chance.

Silk is the only natural filament fibre, traditionally associated with rank or wealth because it has always been expensive and scarce so that it was a status symbol as a clothing fibre.

Cultivated silk is obtained by the *Bombycidae* variety. There are two types of cultivated variety; one is known as *Univoltine* and the other *Multivoltine*. The former gives only one crop of silk during the year is the ordinary European silk worm (*Bombyx mori*). The latter, which gives more than eight crops passing through succession of generations during the years, is the Bengal silkworm. The cocoons of the univoltine silk worms are of a firm and close consistency, so that the silk can be readily reeled off them, and the eggs require a certain degree of cold to hatch out regularly and should be healthy. Whereas the multivoltines hatch out healthy worms without exposure to cold and their cocoons contain a small amount of silk.

The favourable conditions for univoltine variety is prevailing in Kashmir and southern parts of India whereas those for multivoltines prevail in Bengal.

Silk moth lays eggs from which the silk larvae hatch. Larvae are fed on mulberry leaves and spinn cocoons, which are then sorted according to colour and damage.

Wild silk is the non-mulberry feeding variety insects, called Saturniidae. The most important species of silk are (1) Tussar silk, (2) Muga silk and (3) Eri silk.

The chief silk producing countries are China, Korea, Japan, Thailand, India, Uzbekistan, Brazil, Iran, Italy and France. About 1 million workers are employed in the silk sector in China. Silk Industry provides employment to 7.9 million people in India, and 20,000 weaving families in Thailand. China is the world's single biggest producer and chief supplier of silk to the world markets. India is the world's second largest producer. Table 2.4 gives the production of silk in metric tons in different countries.

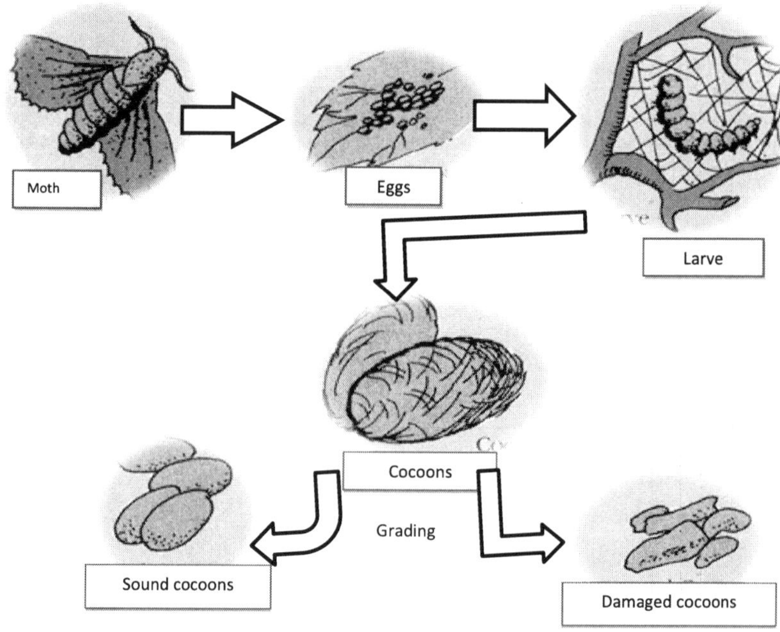

Figure 2.9 Silk process flow.

Table 2.4 Global silk production (in metric tons)

Country	2008	2009	2010	2011	2012	2013
Bangladesh						24
Brazil	1177	811	770	558	614	550
Bulgaria	7.5	6.3	9.4	6	8.5	8.5
China	98,620	84,000	115,000	104,000	126,000	130,000
Colombia	0.6	0.6	0.6	0.6	0.6	0.6
Egypt	3	3	0.3	0.7	0.7	0.7
India	18,370	19,690	21,005	23,060	23,679	26,538

Indonesia	37	19	20	20	20	16
Iran	180	82	75	120	123	123
Japan	96	72	54	42	30	30
North Korea	–	–	–	300	300	300
South Korea	3	3	3	3	1.5	1.6
Philippines	1	1	1	1	0.89	1
Syria	0.4	0.6	0.6	0.5	0.5	0.7
Thailand	1100	665	655	655	655	680
Tunisia	0.08	0.04	0.12	3	3.95	4
Turkey	15	20	18	22	22	25
Uzbekistan	770.5	780	940	940	940	980
Vietnam	–	–	550	500	450	475
Madagascar	15	16	16	16	18	18
Total	**120,396**	**106,170**	**139,100**	**129,662**	**152,846**	**159,776**

The eggs of silk worm are placed in well-lighted and ventilated chambers (one ounce of silk worm eggs yields as many as 36,000 silk worms and about 5 kg of raw silk). After ten days of incubation, the silk worms come out from the eggs. These are fed on chopped mulberry leaves. In just about 35 days after it is born, it grows to 10,000 times its weight at birth. When the caterpillar is about 8 weeks old, it secretes a viscous fluid from the glands in its head. This substance called fibroin is forced through 2 min channels into a side exit near its mouth. At the same time two other glands secrete a gummy liquid called sericin which passes through the same exit. The fibroin, as it emerges from the head of the silk worm, coagulates forming a twin filament coated with sericin. With this twin fine filament the caterpillar produces a protective casing known as cocoon around itself before entering the chrysalis stage of its life (see Fig. 2.9).

In the beginning there are only few fluffy threads by means of which the worm attaches itself to the specially prepared branches of the spinning huts called chandrikas. From here onwards begins the period of transformation. It begins to change into a chrysalis and then into a moth inside the cocoon. The cocoons are stifled by steam or hot air in order to kill the chrysalis within it. The cocoon is ready for reeling.

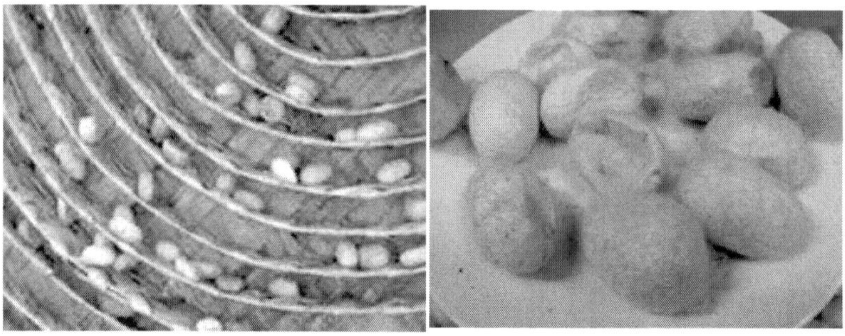

Figure 2.10 Chandrika and cocoons.

In the production of continuous silk filament or nett silk, cocoons are collected and those that are unsuitable for reeling or those intended for supplying the next crop of eggs are removed.

Cocoons are placed in soapy water and threads reeled together to form hanks of filament. They are then degummed and twisted to get silk yarn.

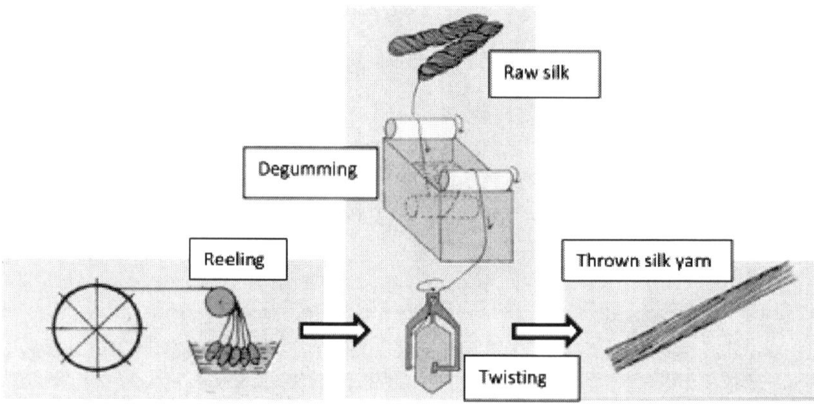

Figure 2.11 Process sequence of silk twisting.

The cocoons sorted for the filature (filatures are the establishments where the silk reeling is done) are treated with hot water in basins and mechanically brushed and then the outer layers of the cocoons are removed. The worker finds the end of the cocoon threads. Nearly 3–8 of these cocoons are reeled together on swift, and the material is wound in the form of a hank.

A fresh end is added to the composite thread immediately when one breaks or runs out from the cocoon.

During its passage to the swift, the filament is dried either by natural or by artificial means. The reeled silk is then taken for twisting. This is called Thrown silk. The twisted silk yarns are wound on spools, or skeins and yarns are now ready for use in weaving. For process sequence, see Fig. 2.11.

Although a cocoon can contain about 2700–4500 m of continuous filament, only approximately 900 m are suitable for filament silk. The remaining is cut into staple and is used in spun silk manufacturing.

Waste filaments from all stages of production of silk are converted into spun silk yarn (Fig. 2.12). This type of material lacks the lustre, fineness and general character of thrown filament silk.

The cocoons are handled in bales of 45 kg. The quality of the silk depends on colour, amount of spinnable fibre on the cocoon and uniformity of cocoons.

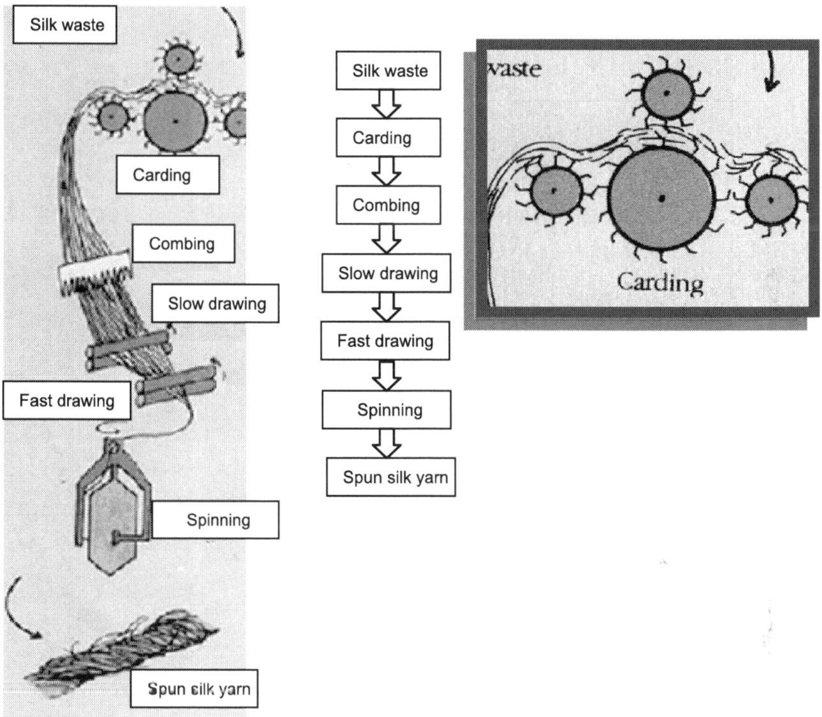

Figure 2.12 Production of spun silk

Physical properties: Silk fibres have a triangular cross section with rounded corners. This reflects light at many different angles, giving silk a natural shine. It has a smooth, soft texture that is not slippery, unlike many synthetic fibres. Its tenacity is 4.5 g/d when dry and 2.8–4.0 g/d when wet.

Silk is one of the strongest natural fibres but loses up to 20% of its strength when wet. It has a good moisture regain of 11%. Its elasticity is moderate to poor: if elongated even a small amount it remains stretched. It can be weakened if exposed to too much sunlight. It may also be attacked by insects, especially if left dirty.

Silk is having a specific strength equal to many engineering materials. It is, in addition, an elastic and resilient fibre similar to wool but not quite as good as wool. This combination of strength and elasticity of silk was unique in the textile fibres until certain manmade fibres were developed.

Unwashed silk chiffon may shrink up to 8% due to a relaxation of the fibre macrostructure. So silk should either be pre-washed prior to garment construction, or dry cleaned. Dry cleaning may still shrink the chiffon up to 4%. Occasionally, this shrinkage can be reversed by a gentle steaming with a press cloth. There is almost no gradual shrinkage or shrinkage due to molecular-level deformation.

Silk is a poor conductor of electricity and thus susceptible to static cling.

Chemical properties: Silk is made up of the amino acids GLY-SER-GLY-ALA-GLY and forms Beta pleated sheets. Inter chain H-bonds are formed while side chains are above and below the plane of the H-bond network. The high proportion (50%) of glycine, which is small, allows tight packing and the fibres are strong and resistant to stretching. The tensile strength is due to covalent peptide bonds. Since the protein forms a Beta sheet, when stretched the force is applied to these strong bonds and they do not break.

Silk is resistant to most mineral acids but will dissolve in sulfuric acid. It is yellowed by perspiration.

Uses: This combination of properties, together with fineness, high degree of lustre, softness and superb drape enables silk to be converted into many beautiful types of fabrics, from delicate chiffons to heavy brocades. The fineness, regularity, strength and elasticity of silk make it suitable for fine screens for printing and parachute fabrics.

Silk's good absorbency makes it comfortable to wear in warm weather and while active. Its low conductivity keeps warm air close to the skin during cold weather. It is often used for clothing such as shirts, blouses, formal dresses, high fashion clothes, negligees, pyjamas, robes, skirt suits, sun dresses and underwear.

Silk's elegant, soft luster and beautiful drape makes it perfect for many furnishing applications. It is used for upholstery, wall coverings, window treatments (if blended with another fibre), rugs, bedding and wall hangings.

It may be interesting to note that a 5 m saree weighing about 400 g will require 1.5 million mulberry leaves weighing 150 kg for the growth of silk worms. This is the quantity of mulberry leaves which 6000 silk worms consume before they spin full cocoons to yield enough raw silk for a 5 m silk saree. In other words, it involves stifling of 6000 silk worms nestling in their cocoons. In fact, however, twice the number of silk worms is required to be reared up because of 50% mortality rate in breeding.

3

Some minor natural fibres

In this chapter some minor utility textile fibres which have not attained commercial status like cotton, wool, silk, jute are discussed.

3.1 Bast fibres

3.1.1 Flax

Though the art of making linen from flax plant was known earlier than 2500 BC, it was only in the 12th century that flax production was organized into an industry. By 1685, AD, Ireland started producing linen fibres from flax plant on a commercial scale. Linen fibre is obtained from the stem of the flax plant. The word linen comes from the Celtic word linen.

Flax (also known as common flax or linseed) (botanical name: *Linum usitatissimum*) is a member of the genus *Linum* in the family Linaceae. It is native to the region extending from the eastern Mediterranean to India and was probably first domesticated in the Fertile Crescent. Flax was extensively cultivated in ancient Egypt. Fig. 3.1 gives the indication of flax production in different countries.

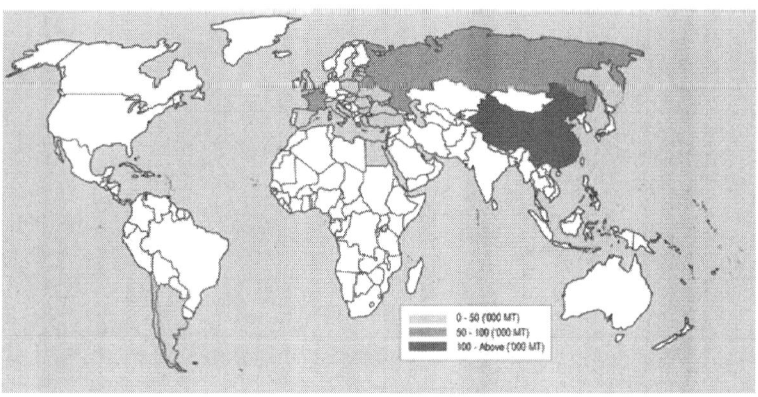

Figure 3.1 Countries growing flax.

The flax plant grows in the temperate climate where there is sufficient moisture. There are two types of flax plants—seed flax and fibre flax. Flax is grown both for its seeds and for its fibres. Fibre flax is optimized for the production of thin strong fibres. Seed flax gives coarser fibres, but far more linseed, since this plant does not have one straight stem, but the stem divides towards various flower heads. Various parts of the plant have been used to make fabric, dye, paper, medicines, fishing nets and soap. It is also grown as an ornamental plant in gardens.

Table 3.1 gives the production of flax in different countries.

Table 3.1 Flax and tow production ('000 MT)

Country	2004	2005	2006	2007	2008	2009	2010	2011
World (Total)	**1013.105**	**1008.656**	**661.958**	**541.087**	**527.851**	**382.538**	**315.583**	**315.084**
China	669.58	695.204	415.265	290.458	257.421	147.51	116.94	126.964
Czech Republic	17.801	13.474	8.757	1.868	0.539	0.461	0.033	0.037
France	95	96	92.4	95	98	74	66.97	52.4
Russian Federation	57.79	55.89	36.11	47.49	52.48	52.56	35.22	43.45
Belarus	56.565	50.43	29.183	38.828	60.894	46.893	45.782	46.034
Belgium	22.215	18.505	16.5	14.17	12.17	13.33	11.52	8.973
Argentina	2.224	2.291	2	2.1	2.032	2.276	2.015	2.323
Bulgaria	0.209	0.2	0	0.1	0.127	0.11	0.088	0.098
Chile	2.4	2.5	2.6	3.292	3.186	2.8	2.479	2.857
Egypt	10	10.5	9	9.514	8.228	8.274	8.3	8.3
Estonia	0.108	0.2	0.1	0.129	0.126	0.082	0.069	0.066
Italy	0.291	0.281	0.278	0.5	0.499	0.494	0.414	0.393
Latvia	1.6	1.2	1.5	1.1	0.2	0.1	0.042	0.3
Lithuania	5.8	3.4	0.3	0.4	0.1	0.031	0.009	0.015
Netherlands	27.4	26.8	22.4	14.2	13.331	14.597	10.519	7.954
Poland	1.439	1.97	0.563	0.502	0.708	0.422	0.197	0.249
Slovakia	0.407	0.28	0.069	0.04	0	0	0	0
Ukraine	16.2	12.7	5.3	3.8	3	1.2	0.4	0.8
United Kingdom	25.596	16.364	18	17.5	14.712	17.294	14.499	13.776

(Courtesy: Ministry of Textiles, Government of India, Powered by IJIRA.)

The flax plant is an annual plant which reaches a maximum height of about 100 cm. The stem is slender and straight, and the flowers are of a pale blue colour. As the fibres for the linen grow in the bast part of the stem of the flax plant, they are designated as bast fibres.

The plants are pulled by hand or machine and care is taken that entire fibre from its top to root is intact. After pulling, the stem part is taken for retting process. In retting the fleshy part of the stem is rotted nearly for a week by contact with water. This process is carried out by exposing it to the action of running or stagnant water or the action of dew. In retting process due to fermentation the gummy substance which binds the fibre to stem is removed. Finally, the flax stem is pressed and beaten against a sharp edge where fibres are loosened. This is called scutching process. Linen fibres of long length are collected at the end of the scutching process. The fibres are collected in bundles and are then dried in the fields. Let us understand the process in detail.

Retting flax: There are several methods of retting flax. It can be retted in a pond, stream, field or a tank. When the retting is complete the bundles of flax feel soft and slimy, and quite a few fibres are standing out from the stalks. When wrapped around a finger the inner woody part springs away from the fibres.

Pond retting is the fastest. It consists of placing the flax in a pool of water which will not evaporate. It generally takes place in a shallow pool which will warm up dramatically in the sun; the process may take from only a couple of days to a couple of weeks. Pond retted flax is traditionally considered lower quality, possibly because the product can become dirty, and easily over-retts, damaging the fibre. This form of retting also produces quite an odour.

Stream retting is similar to pool retting, but the flax is submerged in bundles in a stream or river. This generally takes longer than pond retting, normally by two or three weeks, but the end product is less likely to be dirty, does not stink as much, and because the water is cooler it is less likely to be over-retted.

Both pond and stream retting were traditionally used less because they pollute the water used for that process.

Field retting is laying the flax out in a large field, and allowing dew to collect on it. This process normally takes a month or more, but is generally considered to provide the highest quality flax fibres, and produces the least pollution.

Retting can also be done in a plastic trash can or any type of water tight container of wood, concrete, earthenware or plastic. Metal containers will not work, as an acid is produced when retting, and it would corrode the metal. If the water temperature is kept at 80°F, the retting process under these conditions takes 4 or 5 days. If the water is any colder it takes longer. Scum will collect at the top and an odour is given off like in pond retting. Now enzymatic retting of flax is widely employed.

Dressing the flax: Removing of the straw from the fibres is termed as dressing. It consists of three steps: breaking, scutching and heckling. The breaking breaks up the straw, then some of the straw is scraped from the fibres in the scutching process, then the fibre is pulled through heckles to remove the last bits of straw.

The dressing is done as follows:

Breaking: The process of breaking breaks up the straw into short segments. To do it, take the bundles of flax and untie them. Next, in small handfuls, put it between the beater of the breaking machine (a set of wooden blades which mesh together when the upper jaw is lowered—it looks like a paper cutter but instead of having a big knife it has a blunt arm), and beat it till the 3 or 4 in. that have been beaten appear to be soft. Move the flax a little higher and continue to beat it till all is soft, and the wood is separated from the fibre. When half of the flax is broken, hold the beaten end and beat the rest in the same way as the other end was beaten, till the wood is separated.

Scutching: In order to remove some of the straw from the fi bre, it helps to swing a wooden scutching knife down the fi bres while they hang vertically, thus scraping the edge of the knife along the fi bres and pull away pieces of the stalk. Some of the fi bres will also be scutched away; this cannot be helped and is a normal part of the process.

Heckling: In this process the fi bre is pulled through various different sized heckling combs. A heckle is a bed of "nails"—sharp, long-tapered, tempered, polished steel pins driven into wooden blocks at regular spacing. A good progression is from 4 pins/in^2 to 12, 25, 48, 80 pins/in^2. The fi rst three will remove the straw, and the last two will split and polish the fi bres. Some of the fi ner stuff that comes off in the last hackles is called "tow" and can be carded like wool and spun. It will produce a coarser yarn than the fi bres pulled through heckles because it will still have some straws in it.

Linen fibres are stiff by nature. Because of this linen fibre, fabrics wrinkle readily. Linen fibre fabrics are smooth and possess good feel. It does not soil readily and has a high resistance to biological attack.

Linen fibre thread is used for stitching aircraft and railway carriage upholstery, carpet, suitcases, life belts, tarpaulins, foot balls and cricket balls. Linen fibre is also used in making handkerchiefs, mail bags, hosepipe covers, insulation in telephones, telephone switch boards, parachute harness, table and bed linen, interlining and clothing including pipeline clothing against radioactivity.

3.1.2 Hemp

Hemp is also one of the bast fibres known to ancient Asians, long before the birth of Christ. Hemp is grown chiefly in the Philippines Island, China, Mexico, Russia, West Indies and India. Hemp is a commonly used term for high-growing varieties of the *Cannabis* plant and its products, which include fibre, oil and seed. Hemp is refined into products such as hemp seed foods, hemp oil, wax, resin, rope, cloth, pulp, paper and fuel. Other variants of the herb *Cannabis* are widely used as a drug, commonly known as *marijuana.* Because of its relation with marijuana; there are restrictions for growing hemp in different countries.

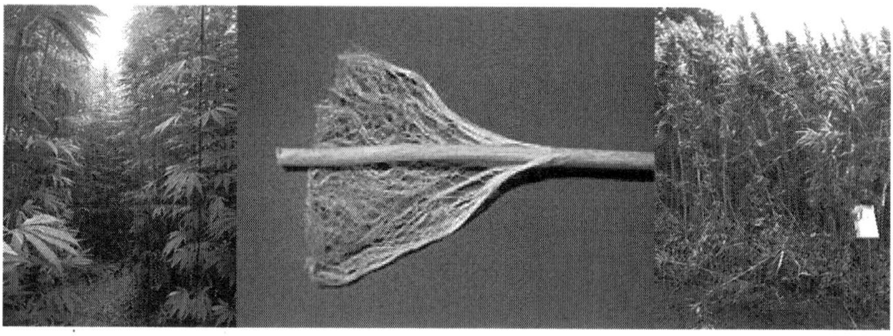

Figure 3.2 Hemp plant and hemp fibre.

In India, Deccan hemp is grown both as crop and hedge plant. It is cultivated largely in Maharashtra, Tamil Nadu and north Gujarat.

The fibres are removed from hemp plant in a manner similar to the process described in flax fibre production. Hemp fibre is a lustrous fibre, has characteristic nodes and joints of linen, but the central canal is wider. The cells are blunt ended, when the fibre is viewed under microscope.

Hemp is one of the fastest growing biomasses known, and one of the earliest domesticated plants known. It also runs parallel with the "Green Future" objectives that are becoming increasingly popular. Hemp requires little to no pesticides, no herbicides, controls erosion of the topsoil, and produces oxygen. Furthermore, hemp can be used to replace many potentially harmful products, such as tree paper (the processing of which uses bleaches and other toxic chemicals, and contributes to deforestation), cosmetics and plastics, most of which are petroleum-based and do not decompose easily.

Industrial hemp has many uses, including paper, textiles, biodegradable plastics, health food and fuel. It is stronger than linen and jute fibre; hence it is ideal for making twine, ropes, cables, carpets, canvas, ship cordage, sail cloth, etc. Hemp is used for a wide variety of purposes, including the manufacture of cordage of varying tensile strength, clothing and nutritional products. This can be used in 100% hemp products, but are commonly blended with fabrics such as linen, cotton or silk, for apparel and furnishings, most commonly at a 55%/45% hemp/cotton blend. The inner two fibres of hemp are woodier, and are more often used in non-woven items and other industrial applications, such as mulch, animal bedding and litter. The oil from the fruits ("seeds") dries on exposure to air (similar to linseed oil) and is sometimes used in the manufacture of oil-based paints, in creams as a moisturizing agent, for cooking, and in plastics. Hemp seeds have been used in bird seed mix.

3.1.3 Ramie

Ramie (*Boehmeria nivea*) is another vegetable origin, bast fibre from *Urticaceae* or *nettles* grown chiefly in China and other neighbouring countries. The Chinese variety is often known as China grass or Rhea. There are also varieties known as white ramie and green ramie. Fig. 3.3 shows the areas where ramie is grown.

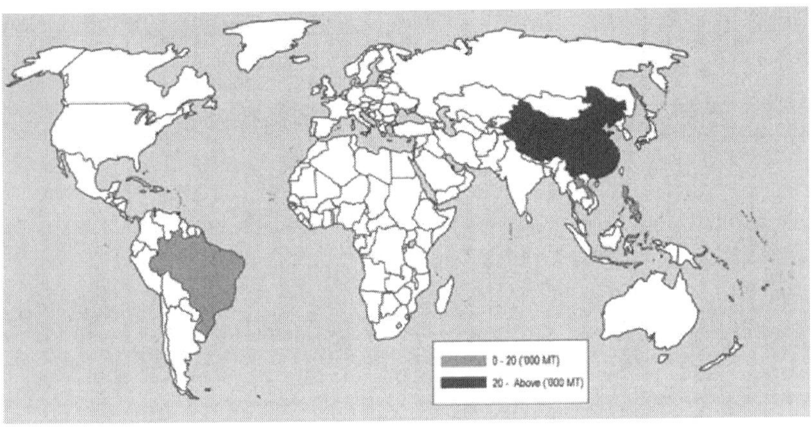

Figure 3.3 World ramie production map.

Table 3.2 Ramie production ('000 MT)

Country	2004	2005	2006	2007	2008	2009	2010	2011
World Total	259.036	281.537	291.545	296.096	255.204	147.892	118.593	128.782
China	255	277.052	286.819	291.259	250.438	143.9	114.08	124
Brazil	1.196	1.158	1.221	1.072	1.023	0.499	0.831	0.971
Philippines	1.57	1.527	1.605	1.765	1.543	1.514	1.53	1.578

Lao People's Democratic Republic	1.27	1.8	1.9	2	2.2	1.979	2.152	2.233

(Courtesy: Ministry of Textiles, Government of India, Powered by IJIRA.)

Ramie is one of the oldest fibre crops, having been used for at least 6000 years, and is principally used for fabric production. It is a bast fibre, and the part used is the bark (phloem) of the vegetative stalks. Ramie is normally harvested two to three times a year but under good growing conditions can be harvested up to six times per year. Unlike other bast crops, ramie requires chemical processing to de-gum the fibre.

Ramie (*Boehmeria nivea*) is a flowering plant in the nettle family *Urticaceae*, native to eastern Asia. It is a herbaceous perennial growing to 1–2.5 m tall; the leaves are heart-shaped, 7–15 cm long and 6–12 cm broad, and white on the underside with dense small hairs—this gives it a silvery appearance; unlike nettles, the hairs do not sting. The true ramie or China grass also called Chinese plant or white ramie is the Chinese cultivated plant. A second type is known as green ramie or rhea and is believed to have originated in the Malay Peninsula. This type has smaller leaves which are green on the underside and it appears to be better suited to tropical conditions.

Figure 3.4 Ramie plant.

Ramie is one of the strongest natural fibres. It exhibits even greater strength when wet. Ramie fibre is known especially for its ability to hold shape, reduce wrinkling, and introduce a silky lustre to the fabric appearance. It is not as durable as other fibres, and so is usually used as a blend with other fibres such as cotton or wool.

It is similar to flax in absorbency, density and microscopic appearance. However, it will not dye as well as cotton. Because of its high molecular crystallinity, ramie is stiff and brittle and will break if folded repeatedly in the same place; it lacks resiliency and is low in elasticity and elongation potential.

Despite its strength, ramie has had limited acceptance for textile use. The fibre's extraction and cleaning are expensive, chiefly because of the several steps—involving scraping, pounding, heating, washing or exposure to chemicals. Some or all are needed to separate the raw fibre from the adhesive gums or resins in which it is ensheathed. Spinning the fibre is made difficult by its brittle quality and low elasticity; and weaving is complicated by the hairy surface of the yarn, resulting from lack of cohesion between the fibres. The greater utilization of ramie depends upon the development of improved processing methods.

Ramie is used to make such products as industrial sewing thread, packing materials, fishing nets and filter cloths. It is also made into fabrics for household furnishings (upholstery, canvas) and clothing, frequently in blends with other textile fibres (for instance, when used in admixture with wool, shrinkage is reported to be greatly reduced when compared with pure wool). Shorter fibres and waste are used for paper manufacturing. Ramie is also used as an ornamental plant in eastern Asia.

Manufacture of ramie fibre is similar to linen fibre process. The finished fibre is fine, silky and strong. This makes it suitable for weaving into fine table-linen like tray cloths, table cloths, napkins, etc.

Ramie has many properties similar to those of linen. A slight amount of gum in the fibre makes starching unnecessary as the fabric will stiffen sufficiently if ironed damp.

Ramie is more absorbent than cotton. Its wet strength is 30–60% more than dry strength. It dries more rapidly than cotton and linen. Ramie has the additional advantages of being unshrinkable, and highly resistant to biological attack.

3.2 Leaf fibres

The leaf fibres are found in the leaves of Monocotyledenous plant. These fibres are generally used for rope making, cordages, mattings and hats.

3.2.1 Abaca fibre

The banana fibre called Abaca or Manila hemp belongs to *Musa sapientum*, *Musa cavendishi* species. It is cultivated as a commercial crop in the

Philippines, Ecuador and Costa Rica. The abacá plant is stoloniferous, meaning that the plant produces runners or shoots along the ground that then root at each segment. Cutting and transplanting rooted runners is the primary technique for creating new plants, since seed growth is substantially slower.

Figure 3.5 Abaca plant, extraction of fibres and products of abaca.

It grows to a height of 3–6 m, forming a duster of stalk. Numerous suckers extend upward from these roots, stalk. Each stalk appears as a tree trunk of 15–30 cm in diameter, expanding to a foliage of blade-like leaves 90–200 cm long and nearly 30 cm wide. This false trunk is built up by the compact overlapping, leaf stems with the greater thickness in the middle of the cross-section.

The stalk contains a pulpy, fibreless central core extending along the entire length of the stalk. The outer sheaths, on the other hand, are shorter, extending only along a portion of the length of the stalk. The position of the sheaths in the stalk determines the colour, texture and the length of the fibre, the outer sheaths are dark brown, or purple green, white in the inner most sheaths.

Fibre extraction is mainly done by two methods. They are hand stripped and machine extracted.

In the hand stripping process the first stalks for fibre production are harvested for 2 to 4 years after the suckers are set out. The entire trunk is cut down and the leaf blades cut off at the top. The defoliated trunks range in weight from 16 to 91 kg. The fibre is located primarily adjacent to the outer surface of the sheath and can be peeled-off readily in ribbons of strips of 5–8 cm wide and 2–4 mm thick, the entire length of sheath. This stripping process is known as tuxying; the strips being called tuxies. Two methods of tuxying are employed in Philippines. In the first method—called Bacnis method, the trunks are pulled apart and the sheath separated according to their position in the stalk. They are then flattened and the fibre is stripped from the stem by cutting the pulpy portion and pulling away the tuxy.

In the second method, the tuxies are pulled off the stalk from one sheath at a time. In either of these methods, the tuxies are tied into bundles of 23–27 kg and brought to the stripping knife for cleaning.

In this process tuxies are pulled under a knife blade which is pressed tightly against the tuxy in order to scrape away the plant tissue between the fibres. The clean fibre is then air-dried and made up into bundles for subsequent grading and baling.

In addition to hand stripping, machines are used where the trunks from which the dark outer sheaths have been removed, are cut into sections of 120–180 cm in length. The sections are then crushed between rolls and the pulpy tissues are scraped away, one half the length at a time, by two large revolving drums, the rims of which are fitted with scrapping blade which scrape the sheath while it is pressed against a bed plate, oven dried, graded and baled.

Most of the banana fibre produced today is used for ropes and cordage, ships cables, etc. It is also widely used for making hoisting and power transmission ropes, wet-drilling cables, fishing nets and liner, and other types of cordage where strength, durability and flexibility are essential. Banana fibre is also used in the manufacture of strong high grade paper, packing cloth for agricultural products, carpets, table mats, handbags, fancy articles, cushioning material, etc.

Banana fibre is multiple celled structures. The lumens are large in relation to the wall thickness. Cross markings are rare and fibre tips pointed and flat, ribbon like individual fibre diameter range from 14 to 50 mm and the length from 0.25 to 1.3 cm, showing the large oval to round lumen.

The Philippines constitutes main sources of banana fibre contributing more than 90% of the total supply. The remaining quantity is produced in North Borneo, Indonesia and various counties of Central America, Asia and Africa.

India has about 500,000 acres under banana plantation, which might yield 1–2 lakhs tons of fibre.

Canton and pacol are also some more varieties of leaf fibres belonging to Musa species.

3.2.2 Sisal

Sisal, with the botanical name *Agave sisalana*, is a species of *Agave* native to southern Mexico but widely cultivated and naturalized in many other countries. Sisal is grown on large plantations in East and West Africa, Java, Mexico and East India. It yields a stiff fibre used in making various products. The term sisal may refer either to the plant's common name or the fibre, depending on the context. It is sometimes referred to as "sisal hemp", because for centuries hemp was a major source for fibre, and other fibre sources were named after it.

The sisal plant is triennial. Sisal plants, *Agave sisalana*, consist of a rosette of sword-shaped leaves about 1.5–2 m (4.9–6.6 ft.) tall. Young leaves may have a few minute teeth along their margins, but lose them as they mature. The leaves grow from the base of the plant, and each leaf is cut by hand close to the ground. It rots readily in salt water. The leaves are beaten by hand and fibres removed. The fibres are washed simultaneously with scraping. The sisal fibres are traditionally used in the manufacture of commercial tying twines, ropes and cord, and have many other uses, including: paper, cloth, wall coverings, carpets and dartboards.

Figure 3.6 Sisal plant.

Sisal fibre is derived from the leaves of the plant. It is usually obtained by machine decortications in which the leaf is crushed between rollers and then mechanically scraped. The fibre is then washed and dried by mechanical or natural means. The dried fibre represents only 4% of the total weight of the leaf. Once it is dried the fibre is mechanically double brushed. The lustrous strands, usually creamy white, average from 80 to 120 cm in length and 0.2 to 0.4 mm in diameter.

Figure 3.7 Sisal fibres.

Sisal fibre is fairly coarse and inflexible. It is valued for cordage use because of its strength, durability, ability to stretch, affinity for certain dyestuffs and resistance to deterioration in saltwater. Sisal is used by industry in three grades:

- The lower grade fibre is processed by the paper industry because of its high content of cellulose and hemicelluloses.
- The medium grade fibre is used in the cordage industry for making: ropes, baler and binders twine. Ropes and twines are widely employed for marine, agricultural and general industrial use.
- The higher-grade fibre after treatment is converted into yarns and used by the carpet industry.

The fibre is also used for non-woven matting, brushing and roving.

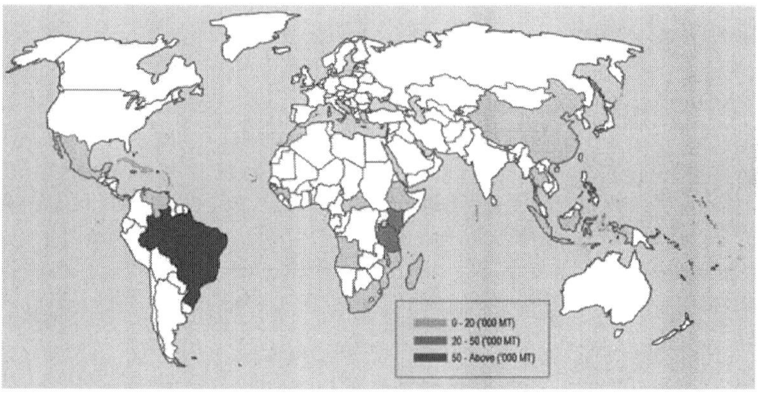

Figure 3.8 World production map of sisal.

Table 3.3 Sisal production ('000 MT)

Country	2006	2007	2008	2009	2010	2011
Africa total (Continent)	69.5	70.6	67.0	52.4	69.9	73.9
Angola	0.5	0.5	0.5	0.3	0.4	0.4
Ethiopia	0.6	0.6	0.6	0.4	0.5	0.5
Kenya	25.5	24.6	22.2	21.5	22.8	28.1
Madagascar	9.2	9.1	8.3	6.9	9.1	8.0
Mozambique	0.6	0.6	0.6	0.4	0.5	0.5
South Africa	1.3	1.0	0.9	0.8	0.8	0.8
Tanzania	30.9	33.3	33.0	21.1	34.8	34.5
Others	1.0	1.0	1.0	1.0	1.0	1.0
Latin America	136.9	129.1	120.9	112.1	115.1	128.4

Brazil	125.5	113.3	105.6	94.9	98.0	111.2
Others	11.4	15.8	15.3	17.2	17.2	17.2
Other countries	60.9	100.3	53.0	46.9	47.3	39.0
China	60.9	100.3	53.0	45.9	46.3	38.0
Total sisal	**267.2**	**300.0**	**240.9**	**211.4**	**232.3**	**241.3**

Courtesy: Ministry of Textiles, Government of India, Powered by IJIRA.)

3.2.3 Pineapple fibre

The pineapple fibre is a soft fibre extracted from the leaves of the pineapple plant and can be produced by breaking the matured leaves through a decorticating machine. It is sometimes combined with silk or polyester to create a textile fabric. Leaves are waste material in pineapple production that can be turned into fibre and sold to contract market. Major pineapple producing countries are Asia – Philippines. It is generally agreed that the pineapple fibres contain one or more of the following matters, i.e. fats, waxes, pectin, hemicellulose, cellulose and lignin. The fibre is extracted either by decortication or by combination of water retting and scutching. The extracted fibre will in the form of long strands and slightly dull yellowish in colour. These fibres are then washed and dried gently combing the fibres in wet condition with five pins to separate the fibres from coarse bundles.

Figure 3.9 Pineapple leaf fibre.

This multicellular fibre is more crystalline. The strength and elongation is comparable with cotton fibre. However, both flexural rigidity and torsional rigidity are higher than cotton fibre. The fibres are mostly used in making ropes, table cloths, mats, handbags, etc. The pineapple fabric can easily be dyed with direct, basic and reactive dyes.

Since piña (Pineapple) is from a leaf, the leaf has to be cut first from the plant. Then the fibre is pulled or split away from the leaf. Most leaf fibres are long and somewhat stiff. Each strand of the piña fibre is hand scraped and is knotted one by one to form a continuous filament to be handwoven and then made into a piña cloth.

Pineapple leaf fibre, which is rich in cellulose, relative inexpensive and abundantly available has the potential for polymer-reinforced composite. Pineapple fibre is considered to be more delicate in texture than any other vegetal fibre. A kilo of leaves may provide up to 15–18 pieces of white, creamy and lustrous as silk fibre about 60 cm long and it easily retains dyes. Fibres of the leaf are scrapped by means of a broken plate or coconut shell and a fast scraper can extract fibre from over 500 leaves per day after which the fibres are washed and dried in the open air. After which they are waxed to remove the entanglements and then the fibres are knotted and bind into yarns for the next process of weaving it into fabric.

3.3 Fruit fibre

3.3.1 Coir

Coir is a versatile natural fibre extracted from mesocarp tissue, or husk of the coconut fruit. It is the natural, cellulosic base, vegetable origin, fruit fibre, obtained from the shell of the coconut.

After thrashing the husk, the long fibre is removed and used for manufacturing a variety of products of multifarious use including home decor, furnishing of luxury apartments and industrial purpose. The coconut husks are softened by steeping them in sea-water and the wood is separated from the fibres by pounding, and then dried. The fibres are about 25 cm in length, these are converted to yarn from which cordage and coarse clothes are prepared, and from the bristles brushes are made.

The unique colour of this fibre has earned for it the favourite name 'the golden fibre although it is known in other countries as coir or cocos fibre'. Products made from coir acquired worldwide reputation and naturally coir industry could attract large markets. This wide acceptance of coir was largely because of its golden sheen, tensile strength and resistance to dampness. These qualities made coir useful in various applications.

Among the traditional industries of the state of Kerala in India coir stands out as the greatest and most important employment giver. Out of the 4 lakhs people employed in this sector 80% are women.

Coir is also widely used in coir mats. Twisted coir is made out of matured brown coir fibre. The products made out of twisted coir are coir mattress, coir cushions, rubberized coir pads, carpet underlays, seat cushions and for insulating drainage pipes. It is picked in coils of 30 kg or in various weights.

Coir mats are made on handlooms, power looms or frames and with or without brush. Creel mats are known for its thin brush. Rod mats for thick brush and fibre mats for compact brush. Latex/rubber backing makes the mats non-slip. Available with woven or stenciled designs and bevelled patterns for use in interior or exterior door fronts.

Coconut shells collected for fibre extraction

Coir being sorted out

Preparing ropes from coir

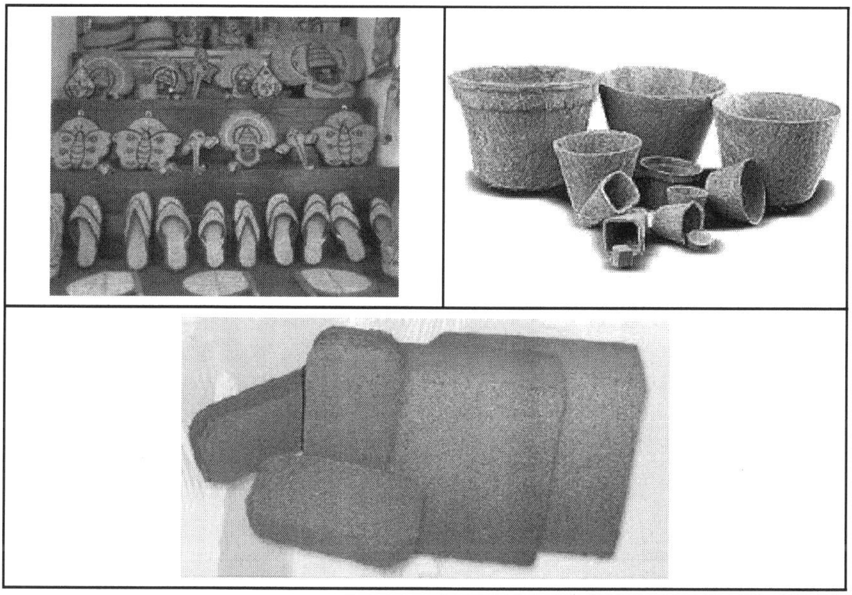

Figure 3.10 Coir and coir products.

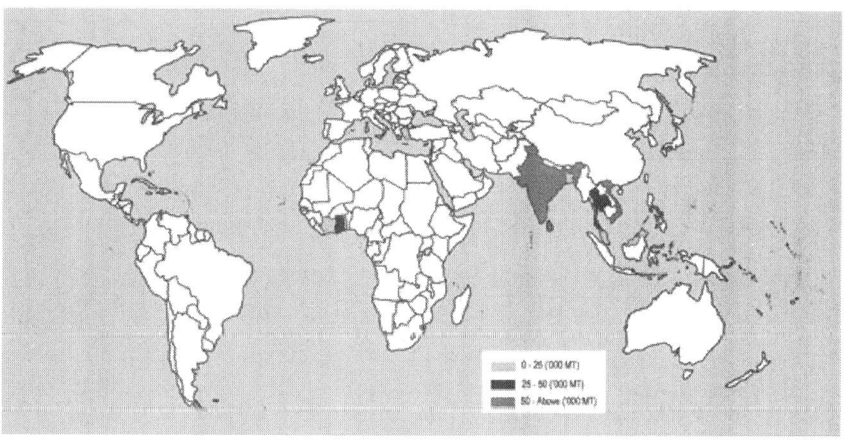

Figure 3.11 World jute production map.

Table 3.4 Coir production ('000 MT)

Country	2006	2007	2008	2009	2010	2011
Brown fibre						
India*	314.0	335.0	340.8	394.0	415.5	425.0
Sri Lanka	80.4	100.1	101.6	169.6	149.8	–

Thailand	41.1	53.7	55.5	64.6	63.7	49.9
Vietnam	91.1	72.2	74.7	91.1	103.0	97.6
Other countries	8.9	11.0	11.0	9.0	9.0	9.0
Total above countries	**535.5**	**572.1**	**583.6**	**728.3**	**741.0**	–
White fibre						
India	96.0	95.0	97.0	97.0	100.0	100.0

(Courtesy : Ministry of Textiles, Government of India, Powered by IJIRA.)

3.4 Mineral fibre

3.4.1 Asbestos

Asbestos is the only natural fibre of mineral origin. When asbestos was first discovered it was called cotton stone by French-Canadians. Asbestos is recovered from the rocks which have crystallized in the fibrous form, was regarded as textile fibre as early as 1950, and was commercially produced in the year 1860. It is found that the technique of spinning and weaving of asbestos was known to Indians as early as 1724.

Asbestos mining began more than 4000 years ago, but did not start large-scale until the end of the 19th century, when manufacturers and builders began using asbestos because of its desirable physical properties, viz. sound absorption, average tensile strength, its resistance to fire, heat, electrical and chemical damage, and affordability. It was used in such applications as electrical insulation for hotplate wiring and in building insulation. When asbestos is used for its resistance to fire or heat, the fibres are often mixed with cement or woven into fabric or mats. These desirable properties made asbestos a very widely used material, and its use continued to grow throughout most of the 20th century until the carcinogenic effects of asbestos dust caused its effective demise as a mainstream construction and fireproofing material in most countries.

$Mg_3Si_2O_5(OH)_4$ chrysotile is one of the six minerals that are regulated as "asbestos". The five others are tremolite, anthophyllte, amosite (grunerite), crocidolite (riebeckite) and actinolite. Out of the six, only two types of asbestos fibres are of commercial interest for the production of asbestos yarn. They are:

a. Tremolite, which are long and strong bluish and lustrous fibres of moderate heat resistance.

b. Chrysolite, long, lustrous, greenish or brown fibres which are weaker than tremolite but are resistant to high temperature.

Figure 3.12 Asbetos stones.

The rocks are crushed and opened to separate the fibres. The fibres are 1–30 cm long with varying diameter. These fibres are carded and spun to get yarn. Asbestos fibres being silicate possess high heat resistance property. Alternatively shorter fibres are mixed with cement to produce asbestos cement heat resistant sheets. These fibres are of little interest from apparel point of view.

		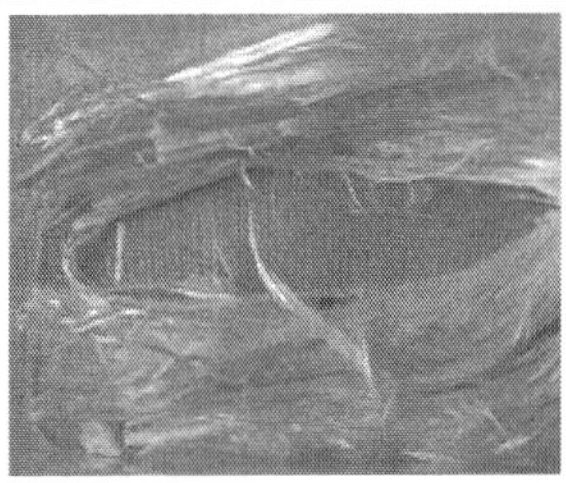
Fibrous tremolite asbestos	Chrysolite asbestos	Asbestos fibre

Figure 3.13 Different types of asbestos.

Asbestos is absorbent and it has wicking ability. Asbestos is resistant to both acid and alkali, therefore, it is used as filters for chemicals and other industrial purposes. It is used for flame-proof clothing of many kinds for laboratory, industrial and military uses. It is used in all types of protective equipment for fire fighting, fire screens, insulation for steam and hot pipes, brake lining insulative building materials, tapes and braids for electrical uses and items wherein non-combustibility is essential. Many times asbestos fibre is used with glass in making decorative fabric for curtains and draperies and for heat insulation.

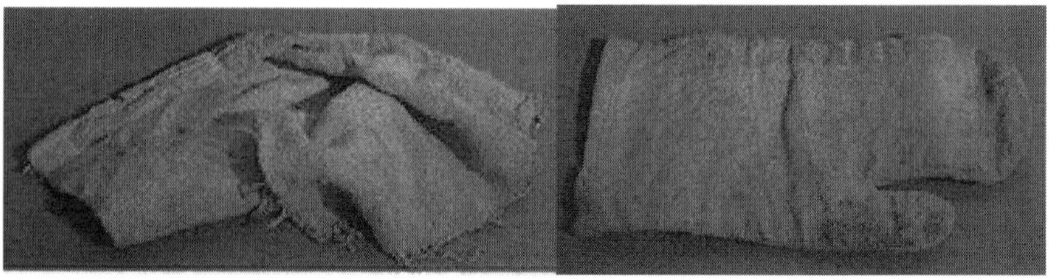

Figure 3.14 Fire resistant fabrics.

4 Manmade fibres

For centuries, man dreamed of producing or imitating the natural silk in order to be able to produce large quantities of the expensive silk fibre easily and cheaply. It was the partial realization of this dream which led to the foundation of the manmade fibre industry. About 100 years ago, fibres in use were all natural fibres.

The early manmade fibres, which include more particularly viscose rayon, cuprammonium rayon and acetate rayon were discovered in the 19th century, since then many varieties of fibres have been added to the list of manmade fibres.

4.1 The position of manmade fibres

In 1950, king cotton was the ruling fibre accounting for nearly 60% of total textiles consumption. Wool accounted for slightly over 20% of consumption; barely ahead of regenerated fibres accounted for 99% of manmade fibre consumption. Consumption of synthetic fibres was in fact much below 10,000 tons.

The consumption of synthetic fibres has made a big leap. It has overtaken wool consumption in 1966, and cellulosic fibres in 1968. Synthetic fibre consumption is now reaching near the consumption of cotton.

The consumption of natural fibres went on increasing until around 1960, when production of synthetic fibres was limited. The huge development of synthetic fibres since that time, going hand-in-hand with a very great drop in prices brought about a decline, in figures of cotton consumption, in absolute figures.

The growth ratio of major synthetic fibres during 1960–1970 comprised 7.5 fold increases in nylon, 14 fold in polyester and 12 fold in acrylic. This increased rate of growth of polyester brought its production almost at par with nylon in 1970. In 1982, the USA and the Western Europe's cotton production accounted for 40% and manmade fibres for 60% and this proportion has more or less stabilized.

In India during the last two decades the share of cotton has come down from 97.5% to about 84.3%. The proportion of manmade fibres has increased by about 6 times. Polyester forms at present only 5% of the consumption of all manmade fibres. This quantum would have been much higher but for the very high excise duty levied. Taking all these factors into account, cotton accounted for 65% while manmade fibres for 35% of the total consumption in the year 2000.

From the above discussion, it is clear that the growth of manmade fibre is truly phenomenal. Now we have reached a stage where the fibre properties can be engineered to suit a particular application or end-use. New regenerated and synthetic manmade fibres are being produced and tested continuously. Manmade fibres have assumed much importance from time to time because of specific quality, such as low cost of production coupled with high performance, which is always a factor in the business world.

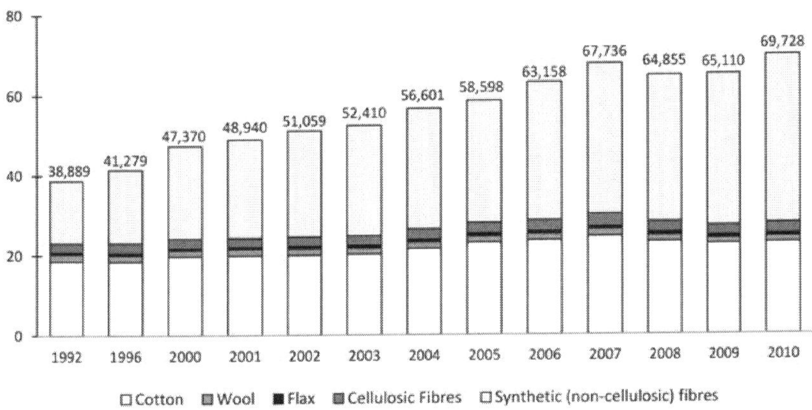

Figure 4.1 World consumption of different fibres.

Table 4.1 The changes in ranking of top consuming countries from 2007 to 2010

Country	Cotton		Wool		Flax		Cellulosic fibres		Synthetic fibres		Total	
	2007	2010	2007	2010	2007	2010	2007	2010	2007	2010	2007	2010
Australia			10	9								
Brazil	7	8										9
China	2	3	1	1	1	1	2	2	1	1	1	1

France		10			5	6	10	10				
Germany	6	7	5	5	10		4	5	6	6	5	5
India	3	2	8	10			3	4	3	3	3	3
Indonesia							6	8		9		
Italy	9				4	5	7	9	9	8	8	
Japan	4	4	2	3	8	10	8	7	4	4	4	4
Pakistan	5	5				8			10		7	7
Poland					6	3						
Russian Federation			9	6	2	2	5	6	5	5	6	6
Saudi Arabia					7	7						
South Korea	10	9	7	8					7	10	10	
Turkey	8	6	6	7			4	3				8
United Kingdom			4	4	9	9			8	7	9	10
United States	1	1	3	2	3	4	1	1	2	2	2	2

Source: Survey 2013 by the International Cotton Advisory Committee (ICAC).]

Table 4.2 World manmade fibre production (billion Lb)

	1990	**1995**	**2000**	**2002**	**2005**	**2010**
Acrylic	5115	5391	5886	5952	5500	5000
Nylon	8241	8244	8976	8609	8500	8150
Polyester	19,131	26,248	42,050	46,200	53,000	64,000
Olefin	6792	9610	12,647	12,076	15,000	20,000
Cellulosic fibre	6079	5371	4884	4670	4450	3900
Other	348	600	691	797	900	1125
Total	**45,706**	**55,464**	**75,134**	**78,276**	**87,350**	**102,175**

Source: Fibre Economics Bureau.

World production of manufactured textile fibre (both synthetic and cellulosic) is projected to rise 5.7% per annum through 2012 to over 62 million metric tons. Polyester will continue to dominate manufactured fibre output. However, the fastest gains will be in specialty synthetic fibre production, such as spandex, aramid and carbon fibres. The Asia-Pacific region, China and India in particular, will continue to be the biggest producer of manufactured fibres, and is also forecasted to experience the most rapid growth.

New fibres

Manmade fibres are not a mere alternative to natural fibres but are new materials of high functionality and high performance. These new fibres can be designed and produced according to the nature of their utilization. In future metals are expected to be replaced by newly developed synthetic fibres of the third generation (see Table 4.3) is superior to metals with respect to strength and modulus.

Table 4.3 Development of fibres over generation

Generation	Fibre availability
First generation fibres	
Before 1950s	Natural fibres
Second generation fibres	
After 1950s	Manmade regenerated and synthetic fibres
Third generation fibres	
1980s	Speciality fibres
1980–84	High performance/high functional fibres
Since 1984	High technology fibres
Fourth generation fibres	
1985	New fibres—Super fibres, smart fibres and nanofibres

As shown in the above table, conventional fibres like cotton, wool, silk, i.e. natural fibres are termed as first generation fibres. With the introduction of manmade fibres (nylon, PET, rayon), era of second generation fibres got started. They offered high performance qualities for fashion fibres as well as offered superior technical properties. For example, the reinforcement in automobile tyres moved from cotton cords in 1900 to a sequence of improved rayons from 1935 to 1955 and then to nylon and polyester which dominate the market now.

Third generation marked the origin of specially fibres—high performance and high functional fibres. These high performance fibres showed a step change in strength and stiffness. They are thus known as high modulus 50–600 gpd (50–600 GPa) and high tenacity 20–40 gpd (3–6 GPa) fibres. These fibres find their application in tyres, composites, aerospace, ballistic sport equipments and many more technical applications. The fibres like Carbon, Aramids, Glass, Teflon, Lycra, Spectra, PBO, Vectran are included in high performance fibres. There is a fourth generation of fibres, which is appearing as the 20th century leads to the 21st century. These are smart fibres and nanofibres which give a new dimension to the use of textiles.

4.2 Advantages of manmade fibres

Manmade fibres are produced in factories, which need not be too far from textile manufacturing areas. Specific qualities of fibres can be produced deliberately and quickly in accordance with the demand.

The filaments can be produced as fine or as coarse as required, staple lengths can be cut exactly to order. Fibres can be produced with high degree of lustre, with reduced or completely dull lustre, as required. Unlike natural fibres, the final product of manmade fibres does not require cleaning.

Most of the fibres are pure white or colourless when produced, but if necessary, colour can be incorporated during production of the manmade fibres.

The controls of quality and quantity that can be exercised in established manmade fibres tend to keep prices steady.

The growth and utility of manmade fibres is mainly influenced by its positive qualities, viz. wrinkle resistance, crease recovery, easy care properties, etc. Manmade fibre fabrics bring out substantial saving on laundry costs; unlike cotton, it can be washed in a basin in the evening, hung up to dry and be worn without ironing the following morning. The light weight characteristic of manmade fibre fabrics gives more mobility because of less weight and quantity—a tourist can take along with, him a few clothes. Hence, these fibres are referred to as Easy-Care fibre fabrics.

Fibres from specific manmade fibre category like homopolymers of vinyl chloride possess a property of negative charge which is said to have outstanding curative and preventing effects on neuralgia and rheumatism. Because of this peculiar property, it is employed in underwears and belly belts.

Most of the synthetic fibres possess high resistance to moth, mildew, insects, mould, which simplify the storage problems, economy of little loss from these causes.

Permanent pleats could be introduced to most of the manmade fibre fabrics because of their unique heat-setting property. Heat-setting property of certain manmade fibres is very useful in texturizing the filament. The extensive development of the knitting industry is to a large extent attributed to the development of textured yarns.

With the increasing knowledge of polymer chemistry synthetic polymer elastic fibres having greater dyeability and flex resistance and higher elastic power were discovered. Those fibres are called Elastomenc fibres.

Apart from the vast industrial and apparel uses, manmade fibres have also a major role in medical uses.

In the recent past, Japanese experts have introduced a newly developed Chameleon fibres based on photochromic materials, which change the colours and colour intensities of the fibre by light exposure (Kamoogawa, Textile Weekly, June 1968, p. 708). According to them when light and water reaching fibres are commercialized, it will be possible to have a swimsuit which may be blue or green when it is wet, but becomes pinkish or orange on drying under the sun.

The day is not far off for the introduction of a fabric which changes the colours and colour intensities depending on the humid environment. Such a new polymer will make profound influence on the growth of the manmade fibre industry.

Introduction of the unique Velcro zipperless zipper-nylon fastener has opened a new horizon in the synthetic fibre utilization. Their versatility is so vast; their application range is limited only by imagination. For instance, collars, cuffs and trims, attached with velcro tapes are quickly removable for cleaning or changing. It saves hours of sewing, eliminates the use of all pins or the strings, botheration of reinforcing buttons, hook and eyes, and snap fastness, provides flawless fit to the wearer.

Nanofibres are exciting new class of fibres used for several value added applications. Special properties of nanofibres make them suitable for wide range of applications from medical to conguron products and industrial to high-tech applications for aerospace capacitors, transistors, drug delivery systems, battery separators, surgery, storage, fuel cells and information technology.

Although the long established natural fibres have excellent and proven qualities, they also have basic properties that cannot be changed and which are, to a certain extent, their limitations.

Manmade fibres have, therefore, been conceived through the desire to make available a wider choice of fibres from which textile manufacturer can select, resulting in a wider choice of fabrics for the consumer (Impact of Manmade Fibres on Wearing Habits and Fashion by H.V. Sreenivasamurthy, *Modern Fibres*, vol. 5, No. I, 1974, p. 30).

4.3 Disadvantages of manmade fibres

Fabrics made from a 100% manmade fibre in 1970–1980 were costlier than the 100% natural fibres as far as India is concerned. This was mainly because of the high cost of raw materials, chemicals and intermediates, plus heavy excise burden levied at every stage. From 1980s onwards increase of petrochemical complexes and relaxing some of the stringent governmental policies resulted in more balanced price.

Some individuals are often prone to skin allergy, because of the dermatological action of manmade fibres. This puts restriction on its use (such problems do not arise in the case of natural fibres).

In general, the manmade fibres are generally hydrophobic in nature; this is necessarily a disadvantage when their products have to be worn next to the skin. These fabrics fail to absorb the perspiration, thus the wearer feels discomfort in hot climate.

Manmade fibre fabrics are a little difficult to sew. Seams do not hold tight as in natural fibre fabrics. So, stitching charges were higher. But this is compensated for by durability and wash and wear properties.

Most of the manmade fibre fabrics manufactured before 1970s failed to provide warmth to the wearer.

In general positive qualities outweigh the negative aspects of the manmade fibres. To achieve improvement or variation in aesthetics, performance and economics, the manmade fibres are blended with natural fibres.

4.4 General principles of manufacturing manmade fibres

The manufacturing process for all manmade fibres is basically the same. The raw materials are treated chemically, and in some cases melted by heating to form a viscous liquid, which is then extruded through very fine holes in a nozzle called spinneret, and the filaments produced are solidified in various ways.

Basically there are three methods of spinning manmade fibres (see Fig. 4.2). They are:

(i) Wet spinning

(ii) Dry spinning Solution spinning

(iii) Melt spinning

Other methods of spinning, viz. gel spinning, electrostatic spinning, etc. are also adopted in spinning of fibres.

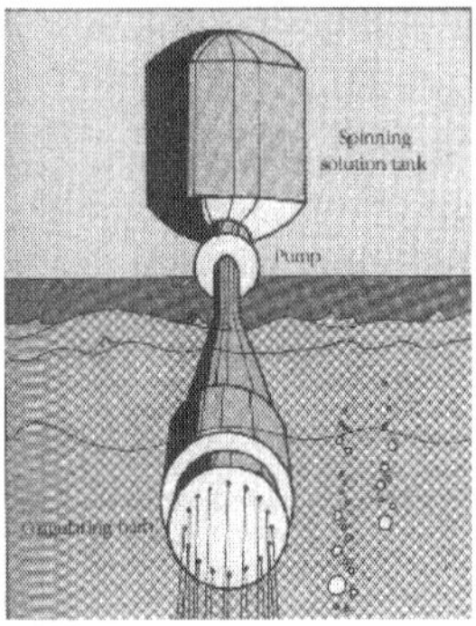

Wet spinning: Rayon, spandex and some acrylic are wet spun

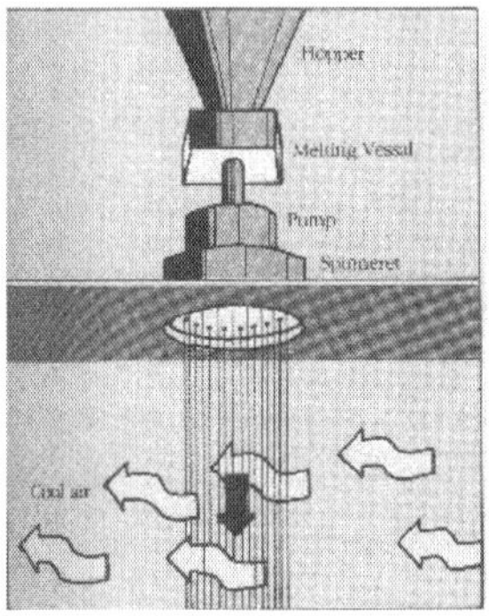

Melt spinning: Nylon, polyester are melt spun

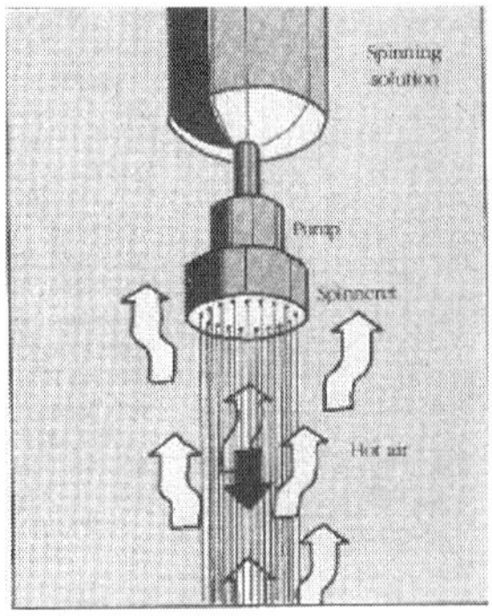

Dry spinning: Acetate, triacetate acrylics, modacrylics are dry spun

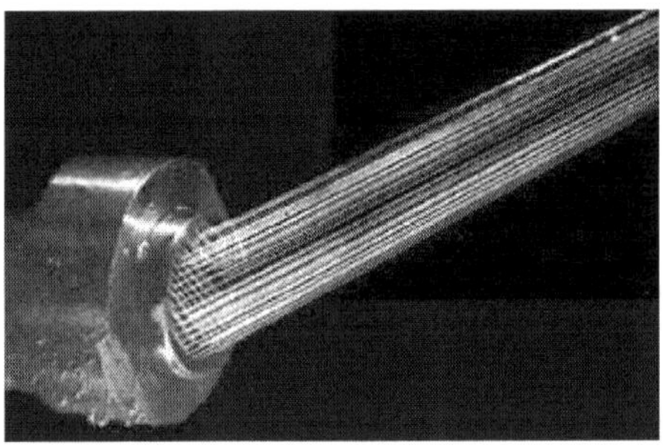

Spinneret

Figure 4.2 Methods of spinning manmade fibres.

Viscose rayon, acrylic, spandex fibres are spun by wet spinning processes. Here the raw material is dissolved by chemicals; fibre is then spun into chemical bath. Fibre solidifies when coagulated by the bath.

Acetate, triacetate, certain of acrylics and mono-acrylic fibres are spun by dry spinning process. Here the resin solids are dissolved in a solvent. Fibre is spun into warm air and the fibre solidifies by evaporation of the solvent.

Nylon, polyester, polyamide, olefin and saran fibres are spun by melt spinning. Here the resin solids are melted in autoclave. Fibre is spun out into the air. Fibre solidifies on cooling.

Wet spinning is a complex process. Washing, bleaching, etc. of the fibre is required before use.

Dry spinning is a direct process. Here a solvent and a solvent recovery plant are required. Washing is not done in this process.

Melt spinning is the least expensive of all the above methods of spinning. It is a direct process. High spinning speed is possible. It does not require washing. Fibres can be shaped by having different shapes of spinneret holes.

These filaments depending on the methods of spinning are drawn and lightly twisted together to form what is called continuous filament yarn. Such yarns are used in weaving or knitting industry.

Many manmade fibres are also manufactured in the staple form. A large number of filaments are collected together (tow) and cut into short staple lengths, and the result is a soft fibrous mass known as staple fibres. These fibres are then combed, drawn, (in other words the tow is converted to top or sliver) and spun into yarn, known as spun or staple yarn, in a similar way to natural fibres. Staple fibres are also blended with each other and with natural fibres before they are spun, to produce blended yarns.

Gel spinning is a type of wet spinning in which polymer solution is kept very dilute to enable low entanglement density. Although, material is having very high molecular weight (one million or more), viscosity is quite high and hence melt spinning is not done. This polymer solution is prepared at a temperature where it is flowable and thus spun through a spinneret. In the solution the molecules become disentangled and remain in that state after the solution is spun and finally extruded to conditions where it forms a gel structure. Because of its sufficiently low degree of entanglement the gel spun material can be drawn to high extent (super drawn). Gel spun fibres are thus ultra strong, high modulus fibres characterized by high orientation and high crystallinity, for example, Spectra, Dyneema (Gel spun polyethylene fibres).

Electrospinning is one of the most popular methods of producing nanofibres. The electrospinning process makes use of electrostatic and mechanical force to spin fibres from the tip of a fine orifice or spinneret. As shown in Fig. 4.3, the spinneret is maintained at positive or negative change by power supply. Electrostatic elongation of polymer threads allows to produce fibres in nano range diameters. When the electrostatic repelling force overcomes the surface tension force of the polymer solution, the liquid spills out of the spinneret and forms extremely fine continuous filaments.

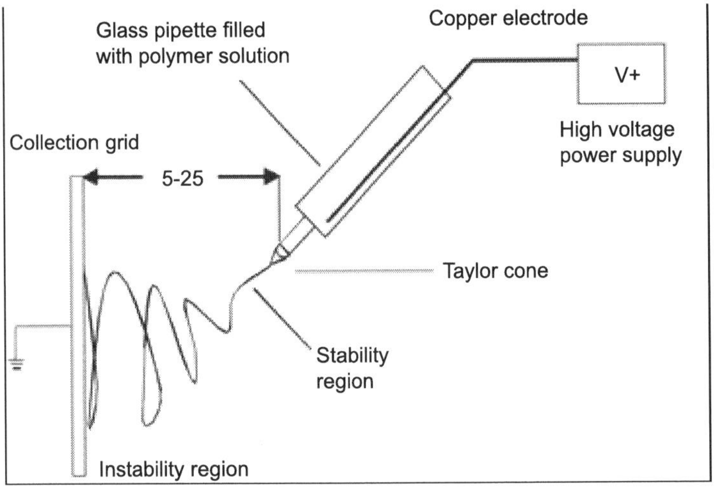

Figure 4.3 Electrostatic spinning.

These filaments are collected onto a rotating or stationary collector with an electrode beneath of the opposite change to that of the spinneret where they accumulate and bond together to form nanofibre web.

The field is generated from high voltage power supply (5–30 kV). The process can be carried out at room temperature unless heat is required to keep the polymer in liquid state. The final fibre properties thus depend on polymer type and other process parameters of electrostatic spinning.

4.5 Nanofibres

Nanofibres are defined as fibres with diameters less than 100 nm. In the textile industry, this definition is often extended to include fibres as large as 1000 nm diameter.

Nanoscience is used to create high strength, lightweight nanofibres with multifunctional properties. These nanofibres have diameter range between 50 and 300 nm (nano microns). There are three methods by which nanofibres are produced: (a) Electrostatic spinning (0.04–2 μm), (b) Melt Blowing Technique (2–10 μm) and (c) Multi-component fibre spinning (15–40μm).

Nanofibres find their applications in various challenging fields. Nanotechnology has been employed to develop fibres possessing a permanent antibacterial performance. Nanofibres can be used in filtration because of their rapid increase in filter efficiency even in filtration of small particles like dangerous bacteria. Nanofibres because of their higher surface area find their application in various innovative fields of textiles. Nanofibres are also used in dyestuff industries for dye desalting and purification. Other uses of nanofibres include producing materials for space shuttles where high strength to mass ratio is required and in medical textiles for tissue engineering where tissues are grown with the help of nanoweb matrix.

The advent of smart nanotextiles will revolutionize the clothes we wear, the furnishings in our homes and the materials used in industry. This coming revolution has heightened the expectations of textile performance, and there is a great demand for "smart fabrics" that are more perceptive of the surrounding environment. Technical and functional textiles may be enlisted in a wealth of applications ranging from military and security to personalized healthcare, hygiene and entertainment.

Advancing the current functionalities of textiles while maintaining the look and feel of the fabric is where nanotechnology is having a huge impact on the textile industry. The market for textiles using nanotechnologies was predicted to reach \$13.6 billion in 2007 and climb dramatically to \$115 billion by 2012. As per the survey of BCC Research published in May 2013, the global market for nanofibre-based products increased from \$128.3 million in 2011 to \$151.7 million in 2012. The market is forecasted to grow at a compound annual growth rate (CAGR) of 30.3% from 2012 to 2017, reaching global revenue of \$570.2 million by 2017.

Textiles, being a pervasive and universal interface, are an ideal substrate for integrating sensors to monitor the wearer and the environment. Textiles offer a versatile framework for incorporating sensing, monitoring and information processing devices. Smart textiles can sense and react to environmental conditions or stimuli, for example, from mechanical, thermal, chemical, electrical or magnetic sources. Some are termed as "passive smart textiles," capable of sensing environmental conditions, whereas "active smart textiles" contain both actuators and sensors, such as thermoregulating garments that maintain the wearer's body temperature. Therefore, the fundamental components within smart textiles are sensors, actuators and control units. The sensing elements, data transmission and processing must be integrated into the textile while retaining the usual tactile, flexible and comfort properties of clothing in order for the smart textile to be practical. Much work in the field of smart clothing features conventional electronics overlaid onto a textile substrate, and the problems of connections, bulkiness, wearability and washability are well documented.

A means of seamless integration is required to develop true textile sensors. This is why nanotechnology is the key to the smart textiles industry, enabling the incorporation of new functionalities at various production stages at the fibre-spinning level, during yarn/fabric formation or at the finishing stage.

The earliest textile developments involved the use of natural materials such as cotton, wool and flax. Later synthetics like nylon, polyester, etc. replaced natural fibres in many technical applications. More recently, synthetic fibres were developed: Lycra®, a segmented polyurethane-urea, has exceptional elastic properties, and Kevlar®, poly-para-phenylene terephthalamide, has ultrahigh-strength properties and is used in bulletproof vests. Today needs for personal mobility, healthcare or rehabilitation require that novel functions in sensing and actuating be integrated into textiles. The fundamental challenge in system-on-textile design is that the drapability and manufacturability of textiles and clothing must remain largely unaffected. Materials suitable for the development of smart nanotextiles include inherently conducting polymers (ICPs), carbon nanotubes (CNTs) and a number of other materials in the forms of nanoparticles or nanofibres.

4.5.1 Inherently conducting polymers

Discovered in 1977, inherently conducting polymers (ICPs) conduct electricity and have the ability to sense and actuate. Actuators based on ICPs can generate much higher stresses with a strain comparable to natural skeletal muscle, and sensors based on ICPs can change their resistivity or generate an electrical signal in response to external stimuli. ICP-based intelligent polymer systems have the ability to sense, process information and actuate.

Most ICPs are prepared via chemical or electrochemical oxidation of the monomer in solution or in the vapour phase, and incorporation of a range of dopants is possible, resulting in ICPs with varying properties. Among the available ICPs, polypyrrole (PPy) is attractive because it has high mechanical strength, relatively stable in air and electroactive in both organic and aqueous solutions. Polyacetylene is unstable in air, limiting its use. Polyaniline (PANi) is one of the most widely studied ICPs, with relatively good environmental stability and good electrical conductivity. PANi has three possible configurations: leucoemeraldine base (fully reduced), emeraldine base (partly oxidized) and pernigraniline base (fully oxidized). When oxidized in aqueous protonic acids, PANi increases its conductivity by 9–10 orders of magnitude. Polythiophene and its derivatives have both *p*- and *n*-type electronic forms, which have been researched to construct polymer field-effect transistors for flexible logic circuits. Moreover, extensive research has also been devoted to its use in polymer solar cells, where its many advantages, such as low production costs, flexibility and light weight, make it suitable for integration into textile fibres.

The unusual electrical conduction in ICPs is achieved through a charged p-conjugated system, which is a truly nano structured charge pathway via a long molecular chain. This pathway is stabilized by a negatively charged dopant incorporated in close vicinity; that is, all charged dopants, such as iodide, triiodide and perchlorate, are spatially removed from the quasi-one-dimensional conduction pathway. The resistive backscattering of electrons is reduced, and this leads to a theoretical conductivity of up to 2×10^7 S cm^{-1}, which is much higher than metal conductors. However, the commonly used bulk synthesis method inevitably results in a shorter molecular chain and introduces conjugation defects, with interchain charge hopping reducing the electrical conductivity of ICPs to less than 1000 S cm^{-1}.

Textiles made from ICPs may be realized by continuous wet spinning to produce ICP-based textile fibres, which can be manufactured into yarns and a range of fabric structures. The conductivity changes in response to external deformation are exploited in the production of textile-based mechanical sensors such as the flexible strain gauge. Particularly, PPy coated on nylon and Lycra by an *in situ* chemical polymerization process has been applied to biofeedback devices for sports training and rehabilitation. Films of ICP nanofibres have been used as sensors to detect chemical vapours that interact with ICPs and change their conductivity. This application exploits the benefit of high surface area resulting from the small diameter of the nanofibres. Thin films made of PANi/CSA nanofibres with diameters between 30 and 50 nm have been used as a chemical sensor with superior performance to vapors of acid (HCl) and base (NH_3).

By incorporating CNTs, the electrical and mechanical properties of ICPs can also be improved. For example, a PANi–CNT composite fibre was produced recently using a wet spinning technique where the ultimate tensile strength and elastic modulus of composite fibre increased by 50–120% with an electronic conductivity of up to ~750 S cm−1. The unique properties of high strength, robustness, good conductivity and pronounced electroactivity of CNTs in the nano domain make these composite fibres potentially useful in electronic textile applications, such as the enhanced force generation when incorporated into fabric as an actuator and the improved conduction when used as the connection wire.

4.5.2 Nanoparticles: composite fibres and finishings

Nanostructured composite fibres are one area where nanotechnology is already having a huge impact within the textile industry. Composite fibres employ nano-sized components such as nanoparticles, graphite nanofibres and CNTs to improve physical properties such as conductivity and antistatic behaviour. Table 4.4 lists some of the nano-sized species that are used to improve the performance of textiles. These nanoparticles may be used to develop composite fibres as nanoscale fillers or through a foam-forming process and may also be applied as fi nishings to the textile, for example, spray-coating TiO_2 for biological protective materials.

4.5.2.1 Applications

A. Sports

The sports industry has driven much research within the textile industry to help improve athletic performance, personal comfort and protection from the elements. Synthetics that were once thought to be inferior to natural fabrics now boast high-performance characteristics. Numerous products designed to improve the comfort of the wearer are commercially available; for instance, there are breathable water-proof fabrics such as Gore-Tex® and moisture-management textiles that wick moisture away from the skin such as Coolmax®. Gore-Tex® fabric uses a membrane of expanded polytetrafluoroethylene (PTFE) that has pores of less than 1 mm in diameter, allowing water vapour to penetrate the material, but preventing the passage of liquid. To maintain the wearer's comfort, it is important that sweat is allowed to evaporate, maintaining the body's natural thermoregulatory function. High-performance moisture-wicking fabrics worn next to the skin, transport perspiration away from the body to the outside of the garment where it can more quickly evaporate. This is achieved using synthetic microfibres that, unlike natural fibres, do not absorb moisture, but rather pass it through by a wicking effect that makes them more comfortable to wear. It is even possible to maintain constant body temperature using phase-change technology such as Outlast Adaptive Comfort®. Phase-changing materials (PCMs) absorb, store and release heat as the material changes phase from solid to liquid and back to solid. A microencapsulation process is used to capture small amounts of phase-change material in a polymer shell so that it is permanently enclosed and protected. These microencapsulated PCMs, developed by Outlast Technologies Inc. and called Thermocules®, can then be applied as a finishing on fabrics or infused into fibres during the manufacturing process.

Through the use of nanotextiles, clothes are adapting to their wearers, meeting their needs and maintaining their comfort. The Adidas running shoe uses sensors, a microprocessor and a motor to adjust its shock-absorbing characteristics to the individual runner's style, pace, body weight and running surface. Nike has also released a smart running shoe containing a wireless sensor that connects to an iPod with various playlists to match the type of workout while also tracking distance, time, pace and calories burned. Textiles are engineered to improve performance; for example, nanotech swimsuits for Olympic swimmers have been developed to reduce drag by using a biometric knitted construction of nylon/elastane with v-shaped ridges that emulate a shark's skin. The latest developments integrate sensing capabilities to provide instantaneous awareness of the physiological condition of the athlete, thus providing valuable information about the athlete's physical abilities, training status, athletic potential and responses to various training regimens. There is a great demand for wearable sensors to be used in the field for kinematic analysis, monitoring of vital signs and biochemical analysis.

Table 4.4 Properties and applications of nanoscale materials used to improve textile performance

Nano-filler	**Properties/applications**
Carbon nanofibres	Increased tensile strength High chemical resistance Electrical conductivity
Carbon black nanoparticles	Improved abrasion resistance and toughness High chemical resistance Electrical conductivity
Clay nanoparticles	Electrical, heat and chemical resistance Block UV light Flame retardant, anticorrosive
Metal oxide nanoparticles (TiO_2, Al_2O_3, ZnO, MgO)	Photocatalytic ability Electrical conductivity UV absorption Photo-oxidizing capacity against chemical and biological species Antimicrobial/self-sterilization
Carbon nanotubes	100 × tensile strength of steel at one-sixth the weight Electrical conductivity similar to copper Good thermal conductivity

Strain sensors made from piezoelectric materials may be used in biomechanical analysis to provide wearable kinesthetic interfaces able to detect posture, improve movement performance and reduce injuries. The conductivity of these textiles is affected by stress and strain applied to the fabric, which can be used to assess physiological movements that impose strain or pressure on the material. Garments integrating piezoresistive ICPs and conductor-loaded rubbers with strain-sensing capabilities offer continuous monitoring of body kinematics and vital signs. The advantage of this approach is that the tactile and flexible properties of the textile are maintained, providing truly wearable fabrics.

Such devices may be used to teach athletes the correct way to perform movement skills by providing real-time feedback about limb orientation.

In addition to their application as strain gauges, conducting polymers have been demonstrated to function as pressure sensors by combining them with compressible textiles. Polyurethane (PU) foam coated with PPy sensors was developed at Dublin City University. The sensor remains soft, compressible, versatile, and, in contrast to conventional coated textiles, is sensitive to forces from all three dimensions. The PPy-coated PU foam has been used for developing a breathing monitor, whereby the foam sample is incorporated into a harness to wrap around the ribcage area. The movement of the ribcage during breathing exerts pressure on the conducting foam, causing an increase in conductivity of the material. It has also been integrated into the armhole of a shirt to detect joint movements of the upper limbs, and smart insoles monitoring plantar pressure have been demonstrated for gait analysis applications.

A new area of research that will have a major impact for sports performance involves integrating chemical sensors into textiles. The aim of the European Union (EU) supported BIOTEX (bio-sensing textile for health management) project is to perform real-time analysis of the various constituents in sweat. Research in this area is lacking, because of the overwhelming focus on blood-based diagnostics. The approach being taken is to integrate electrochemical and optical sensors within a textile substrate, enabling the direct collection of sweat from a large body surface area. The target analytes include sodium chloride and pH, sweat rate and sweat conductivity in addition to monitoring cardiac and respiratory functions. This is of particular interest in sports applications where rehydration strategy plays a critical role in the recovery process after exercise. It is important not only to replace volume losses due to sweat, but also electrolytes. These factors are highly variable among individuals, and current techniques are impractical, involving sweat patches that must be sent to a laboratory for analysis. BIOTEX is developing a wearable system incorporating a fluid handling platform based on moisture-wicking fabrics and nonwoven superabsorbent textiles. The sensing elements are integrated within the fabric's fluidic channels to monitor the sweat composition. Control electronics and wireless data transmission allow real-time analysis of the signal and give feedback to the wearers regarding their well-being, making individuals more aware of their personal healthcare needs.

B. Healthcare

The interest in smart textiles for healthcare arises from the need to monitor patients for extensive periods because of rehabilitation or chronic illness. The problem with conventional clinical visits in these cases is that they can only provide a brief window on the physiology of the patient; wearable devices offer the possibility to monitor physiological signals continuously in a realistic setting. This is vital for the future of the healthcare system, given the global aging population. There is a need to shift the focus of healthcare expenditures from treatment to prevention and wellness promotion.

The EU has funded a number of inter-related, specifically targeted research projects in this area. The WEALTHY (Wearable Health Care System) and MyHeart projects involve wearable textile interfaces integrating sensors, electrodes and connections realized with conductive and piezoresistive yarns to tackle cardiovascular diseases, which are the leading cause of death in the Western world. The WEALTHY system is made up of a sensorized cotton/Lycra® shirt that integrates carbon-loaded elastomer strain sensors and fabric bioelectrodes, enabling the monitoring of respiration, electrocardiogram (ECG), electromyogram (EMG), body posture and movement. Electrodes, to detect ECG and EMG signals, are knitted using stainless steel based yarns, and a hydrogel membrane is applied to improve contact and match impedance with the skin. New products coming onto the market for similar applications include the SmartShirt by Sensatex™ and the Life Shirt® system by VivoMetrics®, offering continuous ambulatory monitoring systems.

There is also potential for monitoring emotional, sensory and cognitive activities, as demonstrated by the MARSIAN system (Modular Autonomous Recorder System for the measurement of Autonomic Nervous system). The system includes a smart glove with sensors for the detection of the activity of the autonomic nervous system, which is responsible for the body's involuntary vital functions. The glove contains noninvasive sensors to measure physiological parameters such as skin temperature, skin electrical conductance and skin potential. A microsensor (0.45 mm) is integrated into the glove to monitor skin temperature, and electrodes measure the skin's electrical activity. The initial approach to electrode integration was to embroider commercially available silver/silver chloride electrodes into a hairnet glove, while a recent prototype uses a 3D structure made of Kapton® copper foil (150 mm thick) with electrodes covered in silver.

One of the more recent endeavours within the EU roadmap is the ProeTEX project (advanced e-textiles for firefighters and civilian victims) to perform on body biochemical sensing within a textile. While BIOTEX fabrics, as discussed previously, monitor the wearer's health, ProeTEX fabrics monitor the surrounding environment to detect any potential risks. The project plans to develop a full system for firefighters and civil protection workers plus a limited system for injured civilians. The wearable sensing and transmission systems will be able to monitor health, activity, position and environment, with information relayed both to the individual and also to a central monitoring unit.

ICPs used for kinesthetic and physiological monitoring, as discussed previously for assessing sports performance, may also be used in the area of patient rehabilitation. These electroactive polymers, typically PANi and PPy, are used as sensing devices and may also be configured as actuators. For this purpose, they are used as electrodes properly configured within an electrochemical cell. By applying a potential, the ICP electrode changes its dimension and works as a mechanical actuator. Integration of such actuators within textiles would enable fabrics to have motor functions, opening a new field of applications, particularly in the development of artificial muscles. For example, ICP actuators have been developed to assist the insertion of cochlear implant electrodes, in which a prototype actuator made of a bilayer PPy actuator is able to steer or bend the electrode in a controllable manner. Although the current actuation force and mechanical energy density of electroactive polymers are relatively low, there is potential to develop rehabilitative aids and orthotic limbs.

Textiles have acted as a second skin for protection and appearance, whereas smart textiles have the potential to emulate and augment the sensory system of the skin by sensing external stimuli such as proximity, touch, pressure, temperature and chemical/biological substances. Lumelsky et al. describe a large-area, flexible array of skin-like sensors with data processing capabilities that can be used to cover the entire surface of a machine, such as a robotic system or even part of a human body. For conditions such as diabetes mellitus, where the patient loses sensation in the limbs, or for bed ridden patients, pressure-sensitive fabrics may aid in assessment and warning to reduce the occurrence of pressure ulcers. PPy foam pressure sensors have been demonstrated for this purpose. With nanotechnologies, smart textiles may provide a touch-sensitive interface/alternative to skin.

Home furnishings may be enlisted into ubiquitous sensing within smart homes for telemonitoring elderly, convalescent or isolated individuals. This aligns with the "continuity of care" concept that wearable technologies bring through monitoring patients at home in comfortable surroundings.

C. Military/Security

There is a need for real-time information technology to increase the protection and survivability of people working in extreme environmental conditions and hazardous situations. Performance improvements and additional capabilities would be of immense benefit to defense forces and emergency response personnel. The SmartShirt by Sensatex™ was initially developed by the Georgia Tech Research Corporation and for military applications. The T-shirt functions like a computer by means of optical and conductive fibres integrated into the garment. The optical fibres are used to detect bullet wounds, pinpointing their exact location and various sensors are used to monitor the body's vital signs during combat conditions.

The Institute for Soldier Nanotechnologies (ISN) is an interdepartmental research centre at the Massachusetts Institute of Technology. The ISN's research mission is to use nanotechnology to dramatically improve soldier survival. The intention is to secure a lighter, faster, more agile force with a heightened awareness of its environment and potential threats. Wireless networking enables medics to monitor the health status of the soldier. The ultimate vision of the battlesuit of the future is a bulletproof jumpsuit, no thicker than ordinary spandex that monitors health, eases injuries, transmits data automatically and enables medics to conduct remote triage of combat casualties to help them respond more rapidly and safely.

The chemical-sensing properties of conductive polymers coated onto woven fabric materials were investigated to detect hazards that may endanger the health of the wearer. Low-ppm detection limits were demonstrated for toxic gases such as ammonia and nitrogen dioxide as well as the chemical warfare simulant dimethyl methylphosphonate (DMMP). Fibre-optic sensors with modified cladding materials are suitable for detecting hazards on the battlefield and may be easily integrated into soldiers' uniforms. The original cladding material is replaced with a chemical agent or environmentally sensitive material on a small section of the fibre. The modified cladding material may be sensitive to different environmental conditions, causing a change in the refractive index. This affects the propagation of the transmitted light signal, which can be measured using optical detection techniques. El-Sherif et al. demonstrated this using a thermochromic agent, segmented polyurethane-diacetylene copolymer, and a photochemical polymer, PANi, as cladding agents.

D. Fashion/Lifestyle

The development of high-tech advanced textiles for specific applications, such as extreme sports, eventually finds its way to street fashion, where designers are allowed the creativity of experimenting with these new emerging materials. Microfibres, for example, were initially developed for space and military applications, and are now used in sportswear, interior fabrics and fashion.

We have become exceedingly reliant on technology; for instance, at any one time, the typical person may be carrying an MP3 player, a laptop computer, a mobile phone, a computational wristwatch and a digital camera. The components of these devices are being continually miniaturized and, with methods such as thin-film technology, the electronics are becoming more flexible. Such advancements are enabling the technology to integrate more easily into our lives and our clothes. Eleksen has developed fabric touch pads integrated into jackets for more accessible control of MP3 players. The initial application was for snowboarding jackets to provide ease of access to the control buttons. Eleksen has also developed a fabric keyboard for personal digital assistants (PDAs) that can be rolled out, easily stored and transported. Another producer of electronic textiles, Textronics, develops fabrics that can warm, illuminate, conduct and sense. One of their recent developments is the NuMetrex heart rate monitoring sports bra, which incorporates conductive knitted sensors that link wirelessly to a heart-rate-monitoring watch.

Nanocoatings now offer advanced protection to improve hygiene and cleanliness. To add antibacterial properties, nano-sized silver, titanium dioxide and zinc oxide are used. Metallic ions and metallic compounds display a certain degree of sterilizing behaviour. It is considered that part of the oxygen in the air or water is turned into active oxygen by means of catalysis with the metallic ion, thereby dissolving the organic substance to create this sterilizing effect. By using nano-sized particles, the number of particles per unit area is increased, and thus, antibacterial effects can be maximized. Antimicrobial coatings are widely applied to socks in order to prohibit the growth of bacteria, but their uses also extend from wound dressings to home furnishings, carpets and clothing. Another innovation for such purposes is superhydrophobic self-cleaning surfaces. This was first inspired by the natural cleanness of plant leaves such as the lotus leaf. Water coming in contact with a superhydrophobic surface forms nearly spherical droplets. Contaminants, either inorganic or organic, adhere to the water droplets and are removed from the surface when the water rolls off. Nano-Tex has a range of products using such coatings to resist spills, repel and release stains, and resist static. These textile enhancements become inherent to the fabric, improving the performance and durability of everyday apparel and interior furnishings.

Although technology may be hidden through invisible coatings and advanced fibres, it can also be used to dramatically change the appearance of the textile, giving new and dazzling effects. Luminex® is a fabric with fibre-optic strands woven into it, which are then, illuminated using light-emitting diodes (LEDs). Luminex® has been incorporated into glowing clothes, safety garments, handbags, furniture and even a wedding dress. Another recent development is the Lumalive fabric from Philips, featuring flexible arrays of coloured LEDs fully integrated within the fabric. These light-emitting textiles can carry dynamic messages, graphics or multicoloured images. Based on concepts of colour and light therapy, brightness and the colour appearance of light are thought to affect mood; these textiles are designed to enhance the observer's mood and positively influence his or her behavior.

Developments in smart nanotextiles may affect many aspects of our daily lives and produce clothing that is contextually aware. New materials integrating novel technologies enable passive, noninvasive sensing of wearers and their environs. A major problem in wearable computing at present is the interconnections, with conventional silicon and metal components being highly incompatible with the soft textile substrate. By integrating technology at the nanoscale, the tactile and mechanical properties of the textile may be preserved, retaining the necessary wearable and flexible characteristics that we expect from our clothing. Smart textiles must be flexible enough to be worn for long periods of time without causing any discomfort in order to become a viable and practical product.

Smart textiles have a large range of applications, often starting as a highly specialized application before becoming a more generally available consumer product. The topics covered here show that this is an area of interdisciplinary research that must involve materials research, sensor technologies, engineering, wireless networking and computer applications. Creating a wearable garment integrates textile and fashion design with input from the end users, such as healthcare workers, defense forces and sports physicians. Market trends suggest great opportunities for nanotechnology within the textile market; given the current pace of development, smart nanotextiles will form a ubiquitous part of our lifestyle. Our clothing is becoming contextually aware and is learning to adjust to suit the individual needs of the user.

4.6 Microfibres

Fibres with thickness less than 0.3 dtex (denier) are termed as microfibres. Microfibres find their use in air filters, dust wipes, etc., because of their high surface area as compared to normal fibres.

The growing demand to increase the fibre properties known to create new sophisticated application for textile materials have been the causes of the rapid growth of microfibre technology. Microfibres are half the diameter of a fine silk fibre, one-third the diameter of cotton, one-quarter the diameter of fine wool and one hundred times finer than human hair. In order to be classified as a microfibre, the fibre must be less than 1 dtex in width. Fabrics made from microfibres are generally light weight, resist wrinkling, have a luxurious drape and body, retain shape and resist pilling. They are also relatively strong and durable in relation to other fabrics of similar weight, and they are more breathable and more comfortable to wear.

A microfibre is defined as a fibre (including staple fibres and filaments) of linear density approximately 1 dtex or less, and above 0.3 dtex. Even finer fibres are produced, of 0.3 dtex or less, but these are commonly referred to as super-microfibres.

Table 4.5 Relationship between fibre linear density and classification

Fibre count, dtex/f	Fibre classification
>7.0	Coarse
7.0–2.4	Medium fine
2.4–1.0	Fine
1.0–0.3	Micro
<0.3	Super-microfibres including nanofibres when their cross-sectional dimensions are within a range of nm, that is of <0.1 dtex or <1 μm

4.6.1 History of microfibres

Japanese fibre manufacturing companies introduced the first 'micro-denier' products during the 1970s. Then followed the developments in Europe during the 1980s, and since the 1990s American fibre manufacturers have been following suit. At present, polyester and nylon are generally used for manufacturing microfibres. However, 'micro-denier' versions of rayon and acrylic products are on the horizon.

For as long as microfibre technology has been around, ultra-microfibre technology has existed as well. These are fibres that are less than 0.3 dtex, and especially within the range of 0.1 dtex. Several different processes can be used to make these fibres, all involving the splitting of a larger fibre into many smaller ones.

4.6.2 Production of microfibres

Microfibres are generally considered to be fibres with a linear density of less than 1.0 dtex. Although the technology for microfibre production has been available for many years now, strong demand for these fibres did not begin until the 1980s. The currently available microfibres are different from ordinary fibres mainly in their dimensions, but have much lower property differences than the standard fibres. Toray was the first company in the world to introduce microfibres, followed by Teijin, Hoechst, ICI, DuPont and others. Recently Toray has introduced an ultra-fine polyester microfibre with a linear density of filament of about 0.05 dtex. This may be called the finest synthetic fibre so far produced commercially.

4.6.3 Manufacturing of microfibres

There are various methods of producing microfibres, including modified conventional spinning. All three conventional spinning methods, namely, melt spinning, wet spinning and dry spinning can be employed to manufacture microfibres. For this method, carefully selected polymerization, polymer spinning and drawing conditions are required. Polyester, nylon and acrylic microfibres may be manufactured by this method. The extrusion spinnerets should contain many holes of very fine diameter, each of which will make one uninterrupted filament, in spite of complex thermal and rheological changes.

4.6.3.1 Dissolved type

Microfibres of this type are manufactured from bi-component fibres with different types of polymers. Comparatively thick bi-component filaments containing different types of incompatible polymers are spun, and the fabric is made using them. When the fabric is treated chemically with solvent, one component is dissolved and removed, and the other component remains as the microfibre. Polyester and nylon microfibres can be made by this method. Commercial production has been reported to use 20/80 ratios of soluble/insoluble polymers to produce a bi-component filament of up to 2 dtex fineness and a final dissolved filament with linear density of about 0.50 dtex. The various combinations of soluble/insoluble polymers reported to form fibres successfully are polystyrene/polyamide and polystyrene/polyester.

4.6.3.2 Split type

The microfibres of this type are obtained by physically or chemically treating the bi-component filaments containing two types of polymers and splitting them into different types of filaments. It is easier to split the segment in filament from itself than in the fabrics. Suitable polymer combinations for splittable bi-component filament spinning are polyamides/polyester and polyester/polyolefines.

Figure 4.4 Cross-sectional photomicrograph of a 16-segment splittable fibre.

4.6.3.3 Direct spun type

This microfibre is directly manufactured by melt spinning. For this method, highly selected polymerization, polymer, spinning conditions and drawing conditions are required.

Special melt spinning dynamics should be considered for the production of low linear density polyester by the direct extrusion method. When polymers have similar dynamic viscosities at the given temperature, the polymer with the lower dynamic viscosity permits the spinning of finer fibres. This result has been attributed to the lower spin-line tension generated when spinning polymers with lower dynamic viscosity. It is postulated that the important parameter in the production of finer PET fibre is the spin-line tension level, which must be kept low in order to obtain finer fibres. The increase in take-up velocity, and the fibre line length between the spinner and the take-up device increase the spin-line stress level, and therefore the minimum fineness attainable increases.

4.6.3.4 Super-drawing technique

In this technique, no molecular orientation is involved. Staple fibre with linear density less than 0.5 dtex can be produced with high drawing ratios. This technique is based on the principle that yarn can be stretched as much as 10–75 times much beyond their conventional draw ratios (3–6 times) if the drawing is carried out at a minimum crystallizing temperature and at special selected drawing conditions, including the temperature range and the type of heating the fibre.

4.6.3.5 Sheath-core spinning method

In this method, two different polymers are mixed, melted and mix-annealed under specified conditions. The conjugate fibre comprising of a concentric circular sheath and a core is manufactured, and the sheath portion is removed to form ultra-fine fibres.

4.6.3.6 Some other methods

- Flash-spinning method
- Solution flash-spinning
- Emulsion-spinning method
- Jet-spinning method
- Centrifugal-spinning method
- Turbulent forming method
- Conjugate-spinning method

4.6.3.7 Bi-component fibres

Bi-component fibres are co-extruded with two different polymers in the cross-section. This allows the fibre to use the properties of both materials, and vastly expands the array of possible fibre performance characteristics. Bi-component fibres are available in staple, filament and microfibre forms. Typical bi-component fibre types are presented in Fig. 4.5.

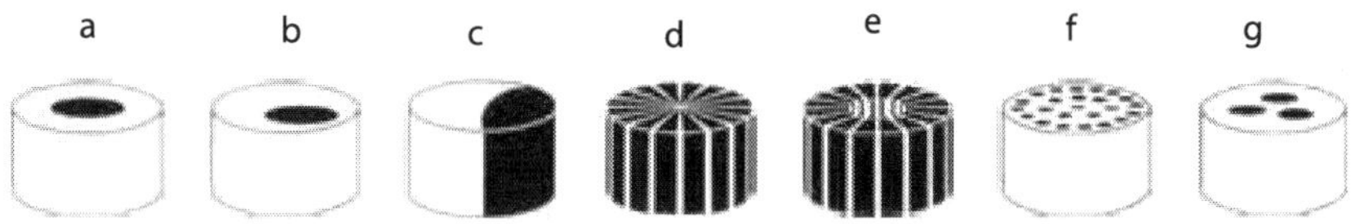

Figure 4.5 Typical bi-component fibre types.

(a) Sheath/core (b) Eccentric (c) Side-by-side (d) Pie wedge (e) Hollow (f) Islands/sea (g) Three islands

4.6.3.8 Specialty cross-sections

A modified cross-section can provide added functionality, such as unique lustre or moisture transport. These cross-sections are available in staple, filament and microfibre forms, and in most cases, are also available as bi-component fibres. Typical microfibres with special cross-section shape, commonly used, are presented in Fig. 4.6; many applications find especially the hollow fibres.

(a) Hollow (b) Trilobal (c) Ribbon (d) 4DG™

Figure 4.6 Typical microfibres with special cross-section shape.

4.6.4 General properties of microfibres

- Ultra-fine linear density (less than 0.1 dtex/f), finer than the most delicate silk.
- Extremely drapeable.
- Very soft, luxurious hand with a silken or suede touch.
- Washable and dry-cleanable.
- Shrink resistance.
- High strength.

- Insulates well against wind, rain and cold.
- Anti-microbial agents help to protect both family members and work staff from the dangers of the bacteria that cause odour and mildew.
- Microfibre is hypoallergenic, and so does not create problems for those suffering from allergies.
- Microfibre is non-electrostatic.
- Microfibres are super-absorbent, absorbing over 7 times their weight in water.
- Microfibre dries in one-third of the time of ordinary fibres.
- Microfibres are environmentally friendly.

4.6.4.1 Comparison between fibres of standard linear density and microfibres

- The tenacity value of fine dpf POY spun is higher than that of normal dpf POY.
- The tenacity value of fine dpf acrylic spun is higher than that of normal dpf acrylic.
- The elongation-at-break value of fine dpf POY spun is lower than that of normal dpf POY.
- The Uster value of fine dpf yarn is slightly higher than that of normal dpf yarn.
- The draw force value of fine dpf yarn is higher than that of normal dpf yarn.

Comparison of the properties of polyester and acrylic microfibres with different linear densities is presented in Table 4.6.

Table 4.6 Physical properties of polyester and acrylic microfibres with different linear densities

Property			Type of fibre			
	Polyester		Acrylic			
Linear density, dtex	1.3	0.8	1.3	0.5	0.5	0.8
Length, mm	38	32	32	40	32	40
Tenacity, cN/dtex	6.32	5.53	3.6	4.92	2.8	3.8
Elongation at break, %	19.5	19.5	32	24.9	20–30	30

4.6.5 Applications of microfibres and micro-filaments

Micro-fibres are characterized by advantageous properties such as pleasant softness of handle, good draping qualities, lustre, bulk and outstanding surface properties. At the start of development, the researchers searched for suitable fields of application for their microfibres, since they had properties which had not yet existed in previous clothing and technical textile concepts.

4.6.5.1 Synthetic game leather and imitation leather

When it was proved that natural game leather collagen fibres have diameters within the range of 4 μm, imitation game leather and artificial leather could be developed with great success, since the new microfibres equated with this level in dimensional terms.

Synthetic game leather and leather products are today produced industrially in Japan by impregnating nonwovens produced from PET, PA or PAN micro-fibres with polyurethane (PU). These products offer outstanding advantages compared to natural leather and game leather in terms of uniformity, dimensional stability, ease of care, colour fastness and low mass.

4.6.5.2 Fashion clothing textiles

Woven fabric was produced from even a 0.1 dtex UFF hollow microfibre combined with a single hollow staple fibre. This product offers softness, bulk, dry handle, a sense of warmth, good recovery and is lightweight.

4.6.5.3 High-performance filter fabrics

Owing to their fine, compact structure, microfibre textiles offer excellent filtration effects for both air and fluid filtration. Independently on common microfibres, also ultra-fine microfibre products, such as 0.05 dtex PP microfibre nonwovens, in combination with a high electrical voltage, which will provide permanent polarization to the nonwoven, attract and absorb charged dust particles.

Micro-fibre textiles can produce excellent fi ltration effects in the process of filtering solid or liquid materials. The characteristics of microfibre liquid filters are as follows: high water passage speed, high extraction performance (retention of particles up to micrometers dimensions) and ease of cleaning microparticles from the filter.

The micrometer-sized fibre segments add mechanical filtration properties to the filter medium. Also, two different dissimilar polymers can be selected that will generate turboelectric properties under flow conditions. The two polymers may also be electrostatically chargeable to enhance first-step filtration properties. These filtration mechanisms, when combined properly, may be able to create a higher efficiency, lower pressures drop filter materials that will last longer before plugging in both air and liquid applications.

Many potential applications exist where synthetic splittable fibres can be used to add value, and create marketing advantages and a head start onto the market for innovative filter suppliers. The addition of a small amount of splittable fibre should increase the dust spot efficiency of filter materials significantly due to its low fibre diameter, in relation to the other fibre diameters currently being incorporated into the filter medium.

In pulsing applications where the filter medium is continuously flexed but also requires stiffness, splittable synthetic fibres add a high degree of reinforcement to the filter medium because the number is at least 16 times the number of fibres available for reinforcement when they are spilt for segmented fibres or more than 33 times for the islands-in-the-sea type fibres.

4.6.5.4 Protection against the weather

Woven sportswear fabrics are also used for protection against wind and weather, and also for insulation purposes. Woven fabrics for protection against the weather were previously coated with polyvinyl chloride (PVC) in most cases.

The PVC coating guarantees absolute waterproofness, but has a serious disadvantage. It allows no passage of air; the wearer perspires after only a few minutes, and has no opportunity of expelling his body moisture to the outside of the garment. This coating is equivalent to an airproof package, and is used nowadays only for heavy duty rainproof clothing (the so-called oil-cloth).

Today, there is a wealth of alternative coatings and methods to replace PVC, such as microporous fluorocarbon coatings, which guarantee some breathing activity on the part of the fabric.

Thanks to ever finer yarns, fabrics can now be produced to meet practically all functional sportswear requirements without additional coating or membranes. They are wind- and water-repellent, yet can breathe.

4.6.5.5 Microfibres for cleaning

Microfibre products are suitable for cleaning anything. Unlike ordinary cleaning fabrics that move or push dirt and dust from one place to another, microfibres actually 'scrap' the dirt or stain from the surface, and then store the dirt particles in the fabric until it is washed. Microfibre cleaning clothes trap dirt and dust inside the cloth, and do not spread dust or dirt around. The user can clean the colthes with water alone no chemicals are needed.

Microfibres are the well-kept secret of professional housekeepers, custodians and car retailers. They are perfect for asthma and allergy sufferers, as they remove dust and dust mites without chemicals. They are also excellent at removing fingerprints from any surface. Grease, tar, splattered bugs come off with the cleaning cloth. They change the static charge on the surface, so it will stay cleaner longer which is important for TV screens, computer monitors and mirrors.

4.6.5.6 Energy conservation

Experiments have shown the drastic savings in the energy consumption in heat exchanges which are brought about by using metal-coated microfibres. In this case, the microfibre shows its heat conveying properties. By using the metal-coated microfibres inside the tubes of heat exchangers, heat transfer enhancement can be achieved.

4.6.5.7 Medical applications

Both manufacturers and consumers are already aware of the many benefits microfibre nonwovens offer to the medical market. When compared the common textiles to microfibre nonwovens, they are lower in cost, easier to use, more versatile, safer and features of better disposability. With this in mind, it is no wonder that microfibre nonwovens are found in hospital surgical drapes and gowns, protective face masks, gloves, surgical packs and bedding and linens.

On the other hand, nanofibres help stimulate living tissues to repair themselves in various parts of the human body, such as cartilage, blood vessels, bones and so forth, due to diseases or wear and tear. Nanofibres are about 10,000 times smaller in diameter than the width of a human hair, but of greater dimensions than a typical cell, which is of the order of 1 μm depending on the type of cell. Nanofibres have a large surface area in comparison to their size. Smaller than usual fibres, they allow cells to adhere to them better, speeding up the healing process.

Another advantage of using nanofibres to grow tissue stem cells from the fact that natural scaffolds themselves are of fibres in the nanometre-size range. In another words, nanofibres are closer replicas of the natural environment of the cells than micron-range fibres.

4.6.5.8 Construction applications

Composites are multilayer materials consisting of different layers with distinct properties in each layer. The composite is constructed in order to utilize a better combination of properties from different layers. Polypropylene and bi-component microfibres can be very important components of fibre-reinforced composites, as they function not only as a reinforcing element, but also as a binder fibre between the individual layers. Polypropylene and bi-component microfibres are used in many different composite products: Microfibre-reinforced concrete (to reinforce and prevent cracks), insulation material (to avoid the use of chemical binders), multifunctional liquid transport media (acquisition and distribution layers), woven fabrics (as a dimensional stability network) and laminated products (lamination between textiles and boards). Polypropylene and bi-component (PP/PE) microfibres have the ability to add structural performance and functionality to the composite materials. Polypropylene and bi-component (PP/PE) microfibres provide the following advantages in fibre-reinforced composites:

- They enable lightweight constructions (PP fibres have the lowest specific gravity of all fibres).
- Easy to process and environmentally friendly thermoplastics.
- Good mechanical properties, toughness and impact strength.
- Stability in rigid environments (resistant to deterioration from chemicals, mildew, perspiration, rot and weather).
- Ability to add bulkiness and softness to the composite.

4.6.6 Some examples of general and spectacular applications of microfibres

- Automotive application to improve air oil filtration, allowing improved engine performance and extended life.
- Perfect for asthma and allergy sufferers, removing dust mites without chemicals.
- Used extensively for hair transplantation, this works to conceal thinning hair.
- Microfibres are also used in sports applications such as sportswear, sports materials, etc.
- Microfibres are extensively used as a luxurious apparel wear.
- Microfibres as swing threads.
- Microfibres for production of synthetic leather.
- Use in the technical textile sector, as well as outside the clothing industry.

- Computer mouse pads, along which the mouse can slide easily, keeping the mouse ball clean at the same time.
- Polishing clothes for wafers and hard disks, acoustic insulation, high performance sound absorption panels and concert hall seat covers, among other products.

4.6.7 Economics aspects of microfibre processing and future prospects for microfibres

The microfibre processing results in:

- 5–10% higher price than conventional fibres.
- Pushing up yarn prices by 4%.
- Reduction in twist by 5–10%.
- 2% net increase in yarn cost.

The microfibres command 5–10% high prices than conventional prices. Microfibres cannot substitute conventional products without significant changes and adjustments to processing methods. This is likely to inhibit the development of microfibres in many textile and industrial applications. In the use of microfibre, the target concept has so far been 'luxury' rather than 'utility', and so the question of costs is perhaps of lesser importance. However, the size of the luxury market is not very large, and with so many companies rushing into the field of microfibre manufacturing, an inevitable price war may bring down the prices of new fibres, obliterating those that are unable to compete effectively. There is also an emphasis on blends, which may have advantages. The question is what ratio of microfibres is required in blends to qualify the product for the microfibre level, and how the proportion of microfibres will cause to pay a premium on price. The industry will have to do the policy in this case.

Microfibres are a completely new generation of ultra-fine synthetic yarns, which have not yet reached their peak of development. There are still a wide range of possibilities to be explored in the design, production, processing and use of this type of fibre. In addition to the common raw materials, PES, PA and PAC, this specialty range will in future include other raw materials such as cellulose.

'Micro denier' yarns have a wider horizon of applications in India due to the traditional dress material designed and made by using natural fibres like cotton and silk. 'Micro denier' yarns have varied simulations of natural fibres, especially silk. We can foresee a brighter future in the Indian market due to the aesthetic properties of the fabrics made out of the yarns. However, the industry has to gear up to handle these delicate yarns in the downstream processes, such as twisting, sizing, warping, weaving, dyeing and finishing.

4.7 Regenerated fibres

For hundreds of years natural fibres have been used for textiles. The production of artificial fibres is of recent origin. A century of research has resulted in the successful manufacture of a number of new fibres. The new fibres have not only out-numbered the natural fibres but also possess many versatile properties. In some of the industrial applications they are found to be indispensable. Silk like fabrics are now within the reach of the common man. More varieties of textiles for clothing are made possible.

Rayon was the first artificial regenerated fibre introduced in the early 19th century. Then started exploration of other fibres with natural raw materials as base, they were named as Regenerated Fibres.

For years scientists pondered over the possibility of simulating the silk by artificial means. The real first step was taken in 1855, when Georges Audemars took Swiss patent for producing nitrocellulose in the form of a thread. The filaments were drawn out by point of needle and then wound onto a bobbin. Seven years later the action of the silkworm was imitated by forcing a solution through fine holes to form a thread.

The present rayon industry can be said to have started in earnest, however, in the year 1878, when three men commenced the experiments which laid the foundation of the work that has continued ever since. They were Count Chardonnet, and the Englishmen Cross and Bevan. Chardonnet concentrated on the production of rayon by nitrocellulose process. In 1892, Cross and Bevan took out their first patent for what they called Viscose silk. A few years later (1904), the problem of suitable means of winding the spun yarn was solved by F. Topham, who invented the rotating yarn box, now called the Topham box.

In 1911, a method of producing cellulose acetate was patented by Dr. Dreyfus. Commercial production of cellulose acetate began in 1921 in England.

This versatile manmade fibre was baptized rayon at quite a later stage. For more than 35 years after it was born, it was known by various names such as artificial silk, art-silk, gloss, wood silk and fibre silk. The word rayon was coined by Kenneth Lord, in 1924 during an industrial sponsored contest.

It is interesting to note that the first rayon factory in India was started in the year 1951. At present there are more than eleven major rayon factories in India.

There are three kinds of rayons produced today for the textile trade. They are: (1) viscose rayon, (2) acetate rayon and (3) cuprammonium rayon.

Out of these three rayons, viscose is the most important from the viewpoint of quantity; more than 80% of the world's production of filament rayon being of this type. Next in order comes acetate with about 16% and cuprammonium with about 4%.

4.7.1 Manufacture of viscose rayon

Viscose rayon is the modern form of one of the earliest types of artificial silk. The name viscose was derived from the word viscous referring to the sticky spinning solution. The name rayon has already been referred to as a generic term desired to replace the word artificial silk.

It is the most versatile and the most widely used manmade fibre. It is the cheapest because it is made from cellulose extracted from wood pulp, a cheap and widely available raw material produced in areas unsuitable for food production.

It is produced in a wide range of filament thickness ranging from the fineness of natural silk to the thickness of coarse wool or hair.

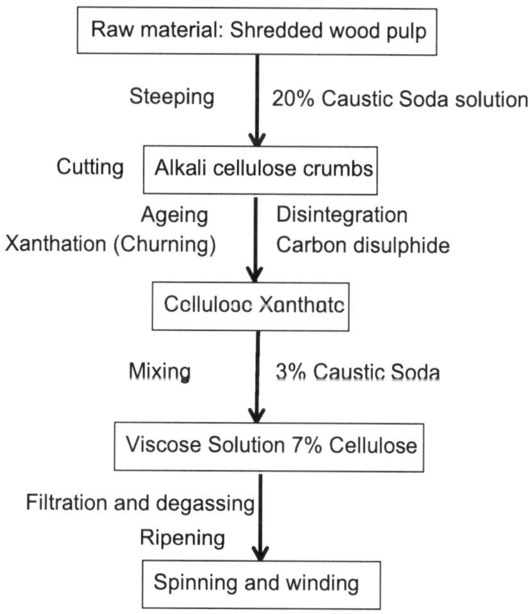

Chart 4.1 Flow chart of viscose rayon fibre manufacture.

The sequence of operations involved in the manufacture of viscose rayon up to the spinning stage may be most conveniently discussed in relation to Chart 4.1 and flow diagram given in Fig. 4.7.

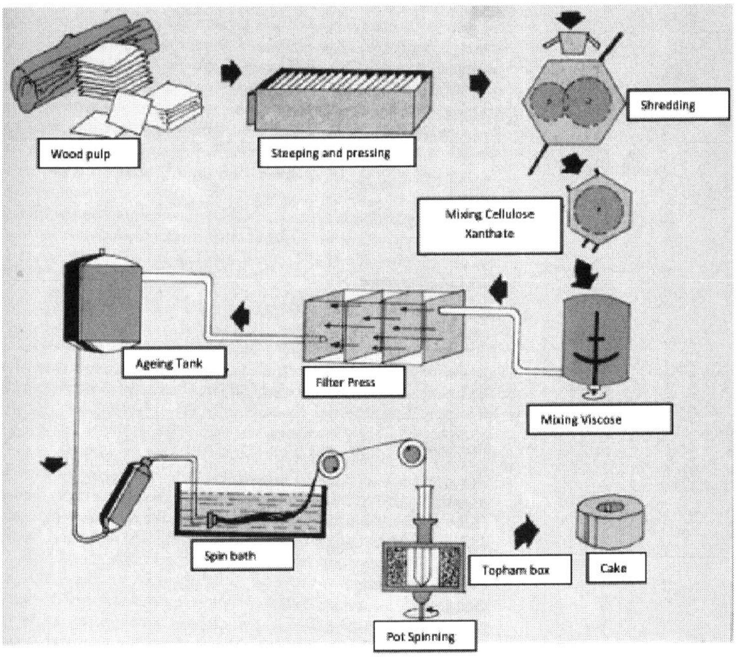

Figure 4.7 Flow diagram of viscose rayon fibre manufacture.

Raw material: The principal source of cellulose used for viscose rayon manufacture is pulp-wood obtained from tree, (especially spruce tree) pulp that has been bleached to free it from gums, lignin and other impurities. The pulp is received in the factory in the form of stiff boards or sheets.

Steeping: It involves treating the pulp boards with caustic soda solution of mercerizing concentration (about 17.5%) to form alkali cellulose. The operation is carried out in a long rectangular tank in which a hydraulic ram is incorporated. Thus when mercerization is complete the excess of caustic liquor can be pressed by the ram, leaving the softened boards of alkali cellulose containing known quantities of caustic soda water. This takes nearly 14 h. The dark brown coloured caustic soda solution is then drained off and the sheets pressed by the movable plates to squeeze the caustic soda solution to a given moist mass of soda cellulose, which is then transferred to a shredding machine for the next operation.

Cutting or shredding: The disintegrating machine or shredder consists of two arms, bearing toothed wings, which work against other similar teeth on the bottom of the vessel. When the vessel rotates, the pulp sheets are cut into small bits called crumbs.

Ageing: The crumbs of soda cellulose are stored in small galvanized steel drums for about 48 h at 23 °C. The time and temperature depend on the nature of the pulp, degree of ageing (maturity) desired, spinning process to be employed, etc.

Xanthation: This process is also called sulphidizing. After the ageing process, the crumbs of soda cellulose are transferred to rotating, air tight chambers. These are double-jacketed, so as to maintain the required temperature during churning. Carbon disulphide is added to the mixer and churned together for 3 h by rotating the mixer at slow speed. Colour of the product changes from white to light yellow, light yellow to deep yellow and finally to reddish orange. Sodium cellulose xanthate formed during this process should not be allowed to form hard lumps as they are difficult to dissolve in the subsequent process. After the churning is over, vacuum is applied for removing the ill-smelling vapours of carbon disulphide from the mixer.

Mixing: The sodium cellulose xanthate, an orange coloured product is taken to a dissolver or mixer, which is provided with stirrers. A diluted solution of caustic soda is also added and the contents are stirred for about 4–5 h, while at the same time, the dissolver is cooled. The sodium cellulose xanthate dissolves to give clear brown, viscous liquor, similar to honey. This solution is blended in big mixers to homogenize any irregularities, and to get more uniform characteristics in the final product.

Filtration and deaeration: This is necessary before the spinning stage is reached, since the rayon is to be formed by forcing the solution through the spinneret having very small holes, all bubbles and solid particles likely to choke the spinneret holes must be removed. Hence, the viscous solution is passed through a series of filter presses on its way to the large tanks in which it is ripened, and later the solution is again filtered several times in the same manner while being transferred from one container to another. Finally, deaeration is accomplished by maintaining a partial vacuum over the viscose solution in the air-tight tanks in which it is kept during the last part of the ripening periods.

Ripening: As the alkali cellulose needs ageing, this viscose solution requires to be ripened to give a solution having the best spinning qualities. Ripening is carried out by storing the viscose solution for 2 to 5 days at 10–18 °C. During this period, a kind of coagulation or throwing down of sediment occurs and the viscosity decreases and then rises to the original value. The sediment necessitates filtration. Hence, the ripened solution is again filtered carefully and deaeratred. The solution is now ready for spinning to produce viscose rayon filaments by wet spinning method.

Spinning: Metered pumps force a measured amount of solution through the multi-holed spinneret, which are arranged in a row, in a long lead tank, horizontally into the coagulating bath.

Spinnerets are designed in a variety of manner to cater to the requirements of wet, dry and melt spinning. Spinneret is a plate which is precision bored with orifices of diameter usually between 0.05 and 0.1 mm (the size of the holes can be varied according to the requirement). Generally, spinneret is made from noble metals like gold, platinum, rhodium or tantalum alloy. During wet spinning the spinneret has to be submerged in liquid baths. Hence, the spinneret which is of thimble shape is required to resist the action of both the liquids, i.e. spinning solution and solutions in coagulating bath.

There is a difference in the spinneret used for wet spinning, dry spinning and melt spinning. For dry spinning and melt spinning spinneret is made of stainless steel or nickel alloy. For melt spinning, spinneret is a flat plate, is of about 6 mm thick and 5–8 cm diameter. The spinneret for filament yarn production has few tens or hundreds of holes. Spinnerets, for staple fibre production, have few hundreds or thousands or in exceptional cases tens of thousands of holes.

The shape of the spinneret orifice determines the cross-section of fibres. It is possible to alter the cross-section of especially melt spun fibres by extruding through a rectangular or triangular hole, by maintaining atmosphere of inert gas at the centre of the orifices. Spinneret holes alone do not control denier of the filament. For example denier of melt spun filament is controlled by output of polymer per hole and winding speed. Denier of filament in wet or dry spinning is controlled by the number of holes in the spinneret, together with stretch ratio.

In the viscose rayon manufacture the wet spinning technique is adopted. The spinneret having many fine holes is submerged in a solution containing 10% sulphuric acid, 18% sodium sulphate, 2% glucose, 1% zinc sulphate and 69% water, and is kept at 40–45 °C in the coagulating bath. Sodium sulphate precipitates or coagulates the dissolved sodium cellulose xanthate and sulphuric acid converts the xanthate into cellulose, carbon disulphide and sodium sulphate. The glucose is supposed to give softness and pliability to the filaments, while zinc sulphate gives added strength.

The quality of viscose rayon filament, thus formed depends on (a) the temperature of the spinning bath, (b) the composition of the spinning bath, (c) the speed of precipitation of the filaments, (d) the period of immersion of filaments in the spinning bath, (e) the speed of the spinning, (f) the stretch imparted to the filaments.

The jet of the viscose spinning solution which enters the coagulating bath through the spinneret is changed into solid filament.

A number of separate filaments are continuously formed and they are steadily drawn away from the spinnerets under a moderate degree of tension, and pass through guide and are collected while being given a small degree of twist in a rapidly rotating spinning pot called Topham box.

Topham box is the take-up device for cake spun in the viscose rayon process. It consists of nearly a cylindrical but slightly tapered in shape, and of about 18 cm in diameter. This is mounted on a spindle, driven by a motor, and rotates at about 7000 rpm.

The filament which comes from tension roller (godet roller) is led through a funnel (as shown in Fig. 4.7) in the top of the box. In the beginning the filaments are thrown outside edge of the pot, gradually cake of rayon is built from the outside to inside towards a hollow centre. A mechanical device provided in the Topham box ensures even build-up of cakes.

The remaining processes include washing the rayon cake with water to remove chemicals, including sulphuric acid, followed by washing with sodium sulphide solution at 62–65 °C to free the filaments from the residual sulphur, the process is called desulphurizing,

If required, the viscose rayon filament yarn may be bleached at this stage with a dilute solution of sodium hypochlorite and then treated with dilute hydrochloric acid, washed thoroughly to remove the acid and finally dried.

4.7.2 Manufacture of polynosic and HWM fibre

Ordinary viscose rayon fibres have low initial wet modulus, i.e. they can be readily stretched by the application of low loads when wet. They have poor elastic recovery, and are very sensitive to the effects of moisture.

Several attempts have been made to improve the performance behaviour of viscose rayon. In 1951, Dr. Shojo Tachikawa of Japan developed a regenerated cellulose fibre called Toramomen which has higher ratio of wet to dry breaking tenacity, increased resistance to swelling by caustic alkalies. Subsequently, other fibre manufacturers entered the field and developed the Toramomen fibre. These fibres are variously known as polynosic, high-wet-modulus (HWM) fibres, etc.

The fibre which is produced by spinning low alkali viscose of extremely high viscosity by low acid coagulation process and which has high degree of polymerization, large crystalline particle size and orientation is called Polynosic fibre. The term polynosic was coined by Dr. M. Drisch of France. It implies multifibril or of high polymerization.

The fibre which is produced by spinning low alkali viscose of comparatively high viscosity in the spinning bath of high zinc containing additives, and which has a fine crystalline particle size and high orientation is called High

Wet Modulus (HWM) fibre. HWM fibres have high tenacity and fairly high wet modulus and higher elongation than that of polynosic fibres.

Polynosic fibre manufacture based on Tachikawa process may be summarized as follows:

Soda cellulose is prepared by steeping wood pulp in 17.5% sodium hydroxide solution, followed by pressing and shredding as in the normal viscose rayon process. The ageing and ripening steps are omitted in this case.

Xanthation is carried out by treating the soda cellulose crumbs with about 70% of carbon disulphide for 2½ h at 15–20 °C. Then the temperature is raised to 25 °C and the dissolution continued for 1 h.

Sodium cellulose xanthate is dissolved in water to give the solution, an equivalent of 6% cellulose and 2.8% sodium hydroxide. In case of ordinary viscose rayon manufacture xanthate is stirred with 3% NaOH to give 7.5% cellulose and 6.5% NaOH.

The spinning solution is spun by extruding through a bath containing 1% sulphuric acid at 25 °C. The filaments are stretched in stages to three times their original length.

Low synthetic acid content (2.5–3%) and low salt concentration (6–7%) in the spinning bath favours slow regeneration of the rayon, slower regeneration and coagulation causes more of stretching. This facilitates the orientation of fibrous molecules to take place before regeneration is complete, with the result high crystallinity can be obtained. It is necessary to control the spinning speed, which should be lower than that in the viscose process. Usually glass spinneret with more holes is used for spinning polynosic fibres,

The conditions described above are typical of those used in polynosic fibre manufacture, but they may be varied in number of ways to provide fibres of desired properties.

4.7.2.1 Comparison of the characteristics of viscose, polynosic and HWM rayons

Rayon is defined as a manufactured fibre composed of regenerated cellulose, as well as manufactured fibres composed of regenerated cellulose in which substituents have replaced not more than 15% of the hydrogen in the hydroxyl groups. Thus, the regenerated cellulose fibres, cuprammonium, viscose, polynosic and HWM fibres are covered under the generic term rayons, while the cellulose acetate fibres in which the hydrogen of the hydroxyl groups are substituted by acetyl groups, and formerly designated as acetate rayon is now excluded and classified separately as acetate. The polynosic and HWM fibres have much in common, and the ISO (International Organization for

Standardization) has now included both the types in the term modal fibre defined as regenerated rayon cellulose obtained by a process giving a high tenacity and high wet modulus.

i Viscose rayon

Viscose rayon is made by converting purified cellulose obtained from wood pulp into xanthate, dissolving the xanthate in dilute caustic soda and then regenerating the cellulose from the product. Wet strength of ordinary viscose rayon is quite low, and this is a great disadvantage. Further, its ability to recover from small deformations is also not good, and this is the reason why viscose rayon fabrics have poor dimensional stability. Being a cellulosic fibre, its chemical properties are similar to those of cotton, particularly with respect to affinity to dyes, etc. However, as considerable breakdown of the natural cellulose molecules takes place during the manufacturing process, viscose is attacked by strong oxidizing agents; acids attack viscose more quickly than they do cotton. While strong solutions of alkali cause swelling with loss in strength. An interesting fact with viscose is that it attains the maximum swelling in 9% concentration caustic soda, while cotton swells at maximum of 18% concentration. This differential behaviour necessitates very careful handling during mercerization of cotton-viscose blends.

ii Polynosic rayon

The polynosic rayon fibres represent a fair attempt to reduce the degree of disadvantage suffered by viscose in relation to cotton. By selecting good quality pulp and reducing the severity of chemical treatments a higher molecular chain-length is preserved (500–700 as compared with 240–270 of viscose rayon) and a microfibrillar structure akin to cotton is obtained by controlled precipitation. The polynosic fibres have a round cross-section, while viscose rayon has a serrated cross-section. The advantages achieved by way of higher tenacity, ratio of wet-to-dry tenacity, wet modulus, etc. can be readily seen in Table 4.7. High wet modulus and good elastic recovery are the most important features of polynosics. They differ from ordinary viscose by having better dimensional stability, ability to withstand mercerization, more crispiness, and lower water imbibition. By and large, the wet processing methods used for cotton could be applied to polynosics.

iii High-wet modulus rayon (HWM)

The high-wet-modulus (HWM) fibres represent the third generation rayon produced by a process differing from the polynosic route, viz. the Toramomen process. The HWM fibres are made by a process which depends on the use of zinc modifier systems; by suitable adjustment of recipes and processing conditions, a wide variation of fibre properties can be achieved starting with fibres of low extensibility, similar to the polynosics, and extending to high extensibility fibres which are less cotton-like in their wet physical properties, but more suitable for blending with synthetic polymer fibres like polyester fibre, where stability can be achieved by resin finishing. The HWM fibres are capable of achieving a high level of crimp that cannot be obtained via the polynosic route; thus, the yarns can be more bulky, and the fabrics more lofty. However, for purposes of blending with cotton, it is desirable to select the low-elongation type of HWM fibre. The high elongation types of HWM fibres could be used with advantage in polyester blends, where the matching of fibre modulus becomes possible, and significant yarn strength improvement obtained over the polyester-cotton equivalent.

Table 4.7 A comparison of physical characteristics of different cellulosic fibres

Sl. No.	**Fibre property**	**Regenerated cellulosic fibre**				**Natural cellulosic fibre**
		Regular viscose	**High tenacity (HT) rayon**	**Polynosic**	**HWM fibre**	**Cotton fibre**
1	Tenacity (gpd)					
	(a) Conditioned	1.2	4.1	4.0	5.0	3.7
	(b) Wet	1.5	3.0	2.8	3.5	4.2
2	Elongation (%)					
	(a) Conditioned	28	30	10	15	10
	(b) Wet	28	37	12	20	13
3	Wet to dry tenacity ratio	60	73	80	70	110
4	Wet modulus at 5% elongation	3	6	21	12	IS
5	Moisture regain (%)	13	12	12	12	7
6	Water imbibition (%)	110	70	60	60	60
7	DP value (degree of polymerization)	175–240	300	300–500	320	5000

Polynosic and HWM fibres represent the achievements made in improving the properties of regular viscose; yet, both types have their advantages and drawbacks. The polynosic fibres have a higher initial wet modulus and alkali resistance than the HWM fibres. On the other hand, the polynosics have rather poor axial and transverse properties, and tend to fibrillate in actual use. The HWM fibres are likely to possess better weaving characteristics. The cost of manufacture is also less for the HWM fibre process. Therefore in the long run, the balance may shift in favour of HWM fibres.

For the last ten years viscose rayon fibre has been recovering fast and consolidating its position in the fibre market following a period of decline, partly attributed to competition and over abundance of synthetic fibre.

Now with the improvements in the manufacturing of viscose rayon process, together with development of improved fibres such as HWM fibres, higher crimp range fibres, model fibres, hollow rayon fibres (Viloft and Avril II), flame retardant fibres (Darelle), alloy rayon fibres, etc., the scope of the viscose rayon fibre has increased tremendously. Viscose rayon is bound to play a very important role in clothing the millions in the years to come.

In the fibre market, manmade cellulosics are seen to have good long-term future as they are produced from renewable resources, i.e. wood pulp from trees which can be cultivated on land that are unsuitable for food production. Furthermore, as cellulosics offers variety, comfort and water absorbancy (see Table 4.8), they are very attractive to the end-users. Cellulosics are also fully biodegradable and the processing of Tencel fibres in newer technology, viz. Dref, Airjet, etc. found to give better performance (see Table 4.9). Tencel is a cellulosic fibre produced by a novel route and with some improved properties. Production of Tencel process utilizes materials which are environmentally clean and recycling of the solvent is an integral part of the process.

Table 4.8 Comparison of fibre properties (cotton with rayon)

Properties	**Cotton**	**Viscose**	**Polynosic**	**Modal**	**Tencel**
1. Titre (den) dtex	–	1.7	1.7	1.7	1.7
2. Tenacity (cNtex)	20–24	22–26	36–38	34–36	40–42
3. Elongation (%)	7–9	20–25	11–13	13–15	14–16
4. Wet tenacity (%)	26–30	10–15	28–30	19–21	34–38
5. Wet elongation (%)	12–14	25–30	12–14	13–15	16–18
6. Wet modulus (at 5% extension)	100	50	210	110	270
7. Moisture regain (%)	08	13	12.5	12.5	11.5
8. Water imbibition	50	90	70	75	65

Table 4.9 Performance of Tencel fibre on yarn properties: A comparison

Yarn property (20 tex)	Type of fibre					
	Tencel		Modal		Cotton (C)	
	Ring spun	Rotor spun	Ring spun	Rotor spun	Ring spun	Rotor spun
Tenacity (gpt)	27.5	19.5	17.5	12.5	15.1	10.2
Elongation (%)	9.4	8.5	8.3	11.6	7.2	5.4
Regularity (CV%)	14.6	17.0	17.2	17.7	15.3	17.2
Neps/1000 m	14.0	30.0	27.0	30.0	98.0	862

Trade names are used within the rayon industry to label the type of rayon in the product. Viscose rayon was first produced in Coventry, England in 1905 by Courtaulds.

Bemberg is a trade name for cuprammonium rayon developed by J. P. Bemberg. It performs much like viscose but has a smaller diameter and comes closest to silk in feel. Bemberg is now only produced in Italy due to United States Environmental Protection Agency regulations in the U.S. The fibres are finer than viscose rayon.

Modal and *Tencel* are widely used forms of rayon produced by Lenzing AG. Tencel, generic name lyocell, is made by a slightly different solvent recovery process, and is considered a different fibre by the US FTC. Tencel lyocell was first produced commercially by Courtaulds' Grimsby plant in England. The process which dissolves cellulose directly in amine oxide was developed by Courtaulds Research in Coventry.

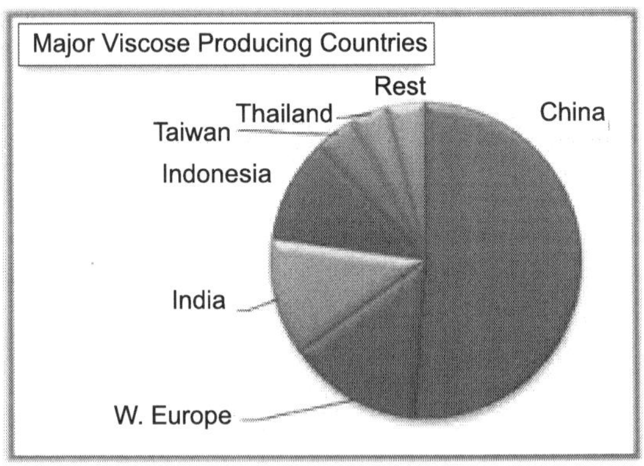

Figure 4.8 Major viscose producing countries.

Source: www.yarnsandfi bers.com for the year 2010.

Galaxy, *Danufil* and *Viloft* are rayon brands produced by Kelheim Fibres. Acordis was a major manufacturer of cellulose based fibres and yarns. Production facilities can be found throughout Europe, the U.S. and Brazil.

Visil rayon is a flame retardant form of viscose which has silica embedded in the fibre during manufacturing.

North American Rayon Corporation of Tennessee produced viscose rayon until its closure in the year 2000.

Grasim of India is the largest producer of rayon in the world (claiming 24% market share). It has plants in Nagda, Kharach and Harihar – all in India, as well as joint ventures in Canada, Laos and China.

iv. Indian varieties

Grasim Industries, the largest producer of viscose and other regenerated cellulose fibres in the world is naturally the leading manufacturer of different viscose varieties in India producing approximately 333,975 tons per annum. In India, various varieties of viscose fibres are manufactured, some of them are:

Viscose plus: Viscose plus brings with it a higher dry and wet tenacity, better dimensional stability and a fuller feel.

High wet modulus fibres (modal): The superior physical properties of Birla modal such as tenacity, elongation and wet modulus make a fabric last longer with an enduring look and feel. Birla modal absorbs moisture more quickly than cotton and displays high dimensional stability.

Polynosic fibres: Polynosic fibre is a new staple with useful physical and mechanical properties. Its tenacity is higher and extension is lower than that of ordinary viscose fibre. It is stable to wet and alkaline treatment and its wet modulus is high.

Birla purocel: Birla purocel is made from a patented zinc free technology making it the only viscose fibre that is totally heavy metal free. The cross section of Birla purocel is specially engineered with less serration and is bean shaped thus influences the desired attributes of softness, bulk and purity. This regenerated cellulose fibre is characterized by a high initial wet modulus of elasticity and a relatively low degree of swelling in sodium hydroxide solution.

Birla excel: Birla excel fibre is the strongest cellulose fibre known and is much stronger than cotton. Birla excel has a distinctive peach skin and leathery feel. Birla excel exhibits high strength and wet modulus that can withstand rigorous processing.

SpunShades is the spun dyed range of Birla cellulose fibres. This is possible as pigments (instead of dyes) are added in the fibre spinning dope itself.

Birla viscose product range also includes antibacterial, chlorine-free and hollow flat super absorbent fibres.

4.7.3 Manufacture of cuprammonium rayon

Cuprammonium rayon is made by a somewhat simpler process than viscose rayon. Like viscose rayon, cuprammonium rayon is also regenerated cellulose. Cotton linters are first subjected to mechanical pretreatment and then to a chemical treatment for justifying cellulose.

The mechanically opened and purified cotton linters are boiled with 2% soda ash solution, to which a little amount of caustic soda may be added. By repeated washing with water, the caustic soda and soda ash may be removed from the cotton. This is followed of the bleaching of the cotton fibres with a dilute solution of bleaching powder and sodium hypochlorite at room temperature for 1 or 2 h. The cotton is then washed, treated with a dilute solution of acid, such as hydrochloric acid and finally washed thoroughly to free it of the acid. The cotton linters are then dried.

The cuprammonium hydroxide solution separately prepared is added to the purified bleached cotton linters, and is dissolved in this solution. The copper-cellulose thus formed may be stored for a long time without the fear of decomposition.

The American Bemberg Corporation is a specialist in the production of these rayons.

In the Bemberg process, the copper-cellulose obtained in the manner explained above, is filtered to remove the impurities. The spinning solution is prepared next in a solution mixer by dissolving the cellulose at a low temperature in aqueous ammonia containing basic copper sulphate. The impurities are again filtered out in a solution filter. The pure, clean spinning solution of dark blue colour and consistency of honey is obtained. The viscous solution is now ready for spinning.

Fact of essential importance is that the cuprammonium solution can be spun at will. It may be spun when the spinning solution is being prepared or it may be used after long storage. The reason is that it does not get decomposed.

Spinning of cuprammonium is done by the methods (i) stretch spinning method and (ii) continuous methods. The difference lies only in post spinning treatments. The filaments are obtained in the same way in both the methods. But in the stretch spinning process the filament yarn is collected on reel or in Topham box, while in modern continuous method it is led in the form of a sheet through washing range. It is then lubricated and dried. The modern continuous method was developed during the wartime for effecting considerable saving as yarn is directly taken up on beam which can be used on warp knitting machine.

In the stretch spinning method, filtered and deaerated cuprammonium cellulose solution is discharged from the nozzle head. The spinneret used is made of nickel and has larger holes (0.8 mm diameter) than those used for viscose rayon manufacture (0.5 mm diameter)

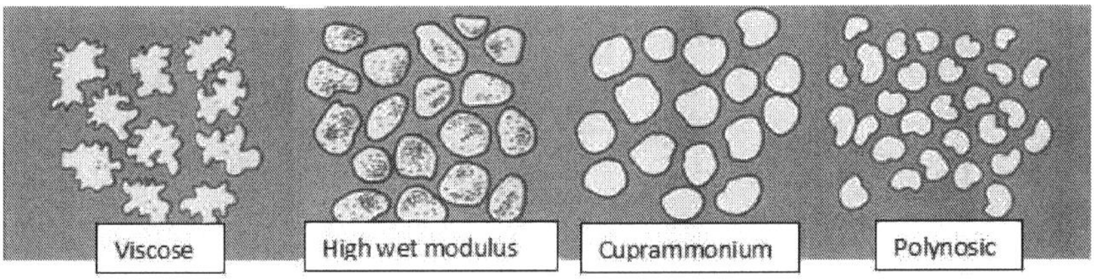

Figure 4.9 Cross-sectional view of rayon.

The spinning solution is extruded through these holes into a long glass funnel as shown in Fig. 4.10. A source of constantly running, soft, (slightly alkaline) warm water is introduced at the top area of the funnel and running out at the bottom. This precipitates the cellulose into plastic form. The filaments on their ways from spinneret to glass funnel is subjected to a considerable amount of stretch, hence the method is called as stretch spinning.

The pull is caused as the filament is drawn-off between steel rod below the funnel and then over the round roller. The steel rod separates the filaments from spinning water, then it is taken into the acid bath which completes the coagulation.

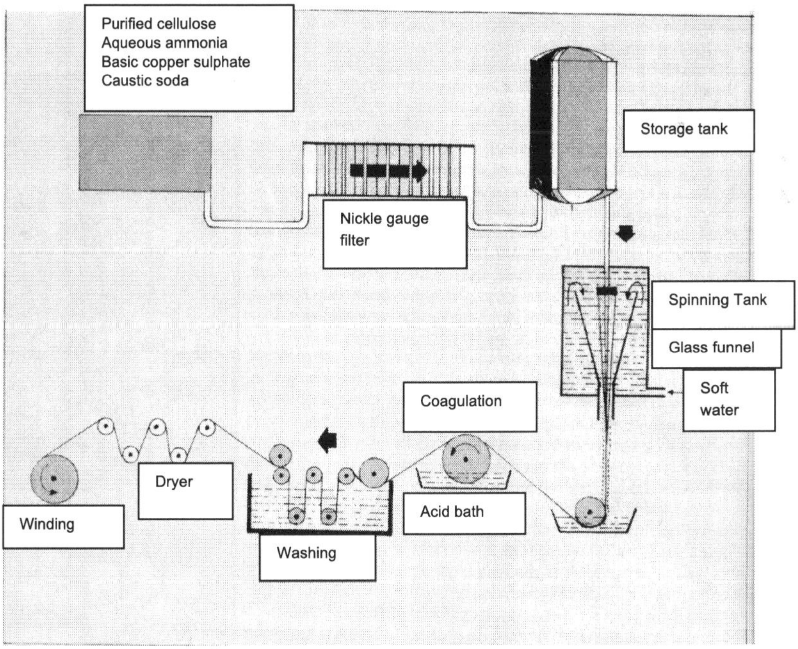

Figure 4.10 Flow diagram showing sequence of cuprammonium rayon filament manufacture.

From the coagulating bath the filaments are wound on bobbins or collected in spinning pot (top ham) or wound on a reel in skein form, or they may be carried on through the continuous method as shown in Fig. 4.10. The wound material is washed to free the filament from acid, copper and ammonia compound, treated with finishing materials according to the requirements, finally centrifuged and dried.

4.7.4 Manufacture of acetate rayon

Cellulose is used as raw material for the manufacture of acetate rayon. This cellulose comes from either cotton linters or wood pulp. Wood pulp is now being used widely for acetate rayon manufacture.

Purification of the raw material itself forms the first stage in the production of acetate, and is done as in the case of cuprammonium rayon manufacture.

The purified cotton linters or wood pulp is kneaded with a mixture of acetic acid, acetic anhydride and a catalyst usually sulphuric acid.

The pretreated cellulose (cotton linters after purification are steeped in glacial acetic acid, which causes swelling, increases the chemical reactivity) is transferred and loaded in a closed reaction vessel. Mixture of acetic acid and acetic anhydride in excess is also added to it. These are thoroughly mixed together. The reaction is hastened by adding sulphuric acid.

Acetylation of cellulose now proceeds. Sulphuric acid reacts with acetic anhydride. This reaction is exothermic and vessel is cooled to maintain the low temperature, otherwise it may cause degradation of the cellulose. For the first 4 h the temperature is kept at 20 °C, it is increased to 25–30 °C for the next 7–8 h. Failure to maintain this range of temperature will lead to breakdown of long chain cellulose molecules into shorter molecules. After 7–8 h the mass becomes gelatinous and very viscous. As a result of this reaction cellulose dissolves in the reaction mixture and it is then considered to be fully acetylated. At this stage three molecules of acetic acid combine with one molecule of cellulose to form cellulose triacetate. This is known as primary acetate.

This primary acetate is not suitable for the manufacture of acetate rayon for the reason that it is soluble only in objectionable and expensive solvents such as chloroform. So hydrolysis treatment is followed to change this.

In the hydrolysis treatment chloroform soluble primary cellulose is converted into acetone soluble secondary acetate.

The solution of primary acetate is transferred to another vessel. It is run into water together with excess of acetic acid anhydride. This results in 95% solution of acetic acid in water. The mixture is allowed to stand for 20 h at high temperature. During this period, hydrolysis of cellulose triacetate takes place. Acetyl groups are replaced by hydroxyl groups. It is actually a partial reversal of acetylation process. Partially acetylated cellulose acetate made in this way is referred to as diacetate. Degree of acetylation in the case of diacetate is about 2.3.

But the secondary acetate used in spinning of acetate is different from diacetate.

Hydrolysis of triacetate is allowed to proceed until 2½ molecules of its acetic acid are combined with one molecule of cellulose. This is known as secondary acetate. This particular cellulose acetate is insoluble in chloroform; but readily soluble in acetone, a solvent which is comparatively cheap and not objectionable from the process point of view.

The secondary acetate is thoroughly washed, centrifuged and dried at low temperature. In batch production, each batch is tested for viscosity and acetyl content. Blending of different batches is always done to ensure uniformity.

Secondary acetate from various batches is mixed in a closed vessel with stirrer with about three times its own weight of acetone. This preparation is called Dope. At this stage addition of titanium dioxide is added for dulling the yarn if so desired.

Well-mixed dopes (dope contains 25–35% cellulose acetate) from different mixers are again collected and mixed in a blending tank. It is filtered, deaerated and then run into a feed tank. This solution has just about the right viscosity to enable it to be spun into rayon filaments.

We have seen that in case of viscose and cuprammonium rayon the cellulose solution is spun into aqueous coagulating liquor. It is quite different with acetate rayon. Acetone being volatile liquid readily evaporates in warm air. So the spinning operation can be made very simple.

The method of spinning adopted is known as dry-spinning. It consists of forcing the cellulose acetate solution through spinnerets within small chambers or cells through which hot air is fed near the bottom at a temperature of 100 °C. This evaporates practically all acetone in dope emerging hum the jets. Acetone laden air is withdrawn from the top. As the acetone evaporates solidified filaments of cellulose acetate are left and are collected through an eye and they are wound on the bobbins at the other end of the chamber as shown in Fig. 4.11.

Since the cellulose acetate is so well purified before it is dissolved in acetone to form a spinning solution, the resulting rayon filaments are equally pure and require no further treatment except for grading according to quality.

Both acetic acid and acetone are sufficiently valuable to warrant the installation of recovery plant so that there is a minimum loss of these chemicals.

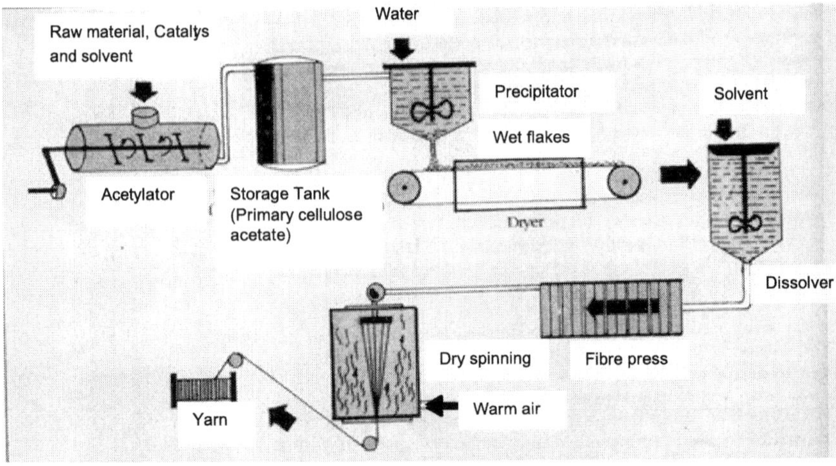

Figure 4.11 Flow diagram showing acetate rayon manufacture.

i. Similarities and differences between cellulose acetate and cellulose triacetate

In the manufacture of cellulose acetate, the cellulose is acetylated in batches in closed mixing tanks equipped with large stirrers. When acetylation is completed, the fibrous structure of the linters disappears, and the charge, as it flows from acetylator, is a viscous, semi-transparent, and acrid-smelling fluid. Cellulose is capable of combining with three molecules of acetic acid to form cellulose triacetate. In other words complete acetylation of cellulose yields cellulose triacetate. This is called primary cellulose acetate.

Cellulose triacetate can be precipitated from the acetylated mixture by dilution with water and then washed free of acid. It is unsuitable for the manufacture of acetate rayon, because it is too impervious to moisture and dyes to permit practical finishing and dyeing.

Cellulose acetate filament yarn must satisfy the following criteria:

i. It must be sufficiently water sensitive, to swell slightly in the dyebath, permitting the fibre to imbibe dye.

ii. It must be sufficiently high in acetyl content to have stability, or resistance, to boiling water.

To produce cellulose acetate that will satisfy the criteria just enumerated, hydrolysis of triacetate is to be carried out. Hydrolysis of triacetate results in an acetyl content which is about half way between the triacetate and the diacetate.

The secondary acetate is run into a large washing vat in which the flakes are allowed to settle without stirring. The clear waste liquor containing dilute acid is drawn off and piped to the acid recovery plant. The cellulose acetate is then washed thoroughly to tree of uncombined acid by stirring the flakes in vat full of fresh water. Finally, the suspension of acetate flakes in water is run into a centrifuge and is dried and delivered to the spinning department.

Some of the chief properties of acetate fibre (Courtaulds-Dicel) are given below:

a. Acetate has particularly rich and attractive appearance.
b. Age has only slight effect on its strength and it will give lasting wear if properly cared for.
c. It has exceptionally good draping qualities.
d. It has good dyeing and printing properties.
e. It has good natural recovery from creasing.
f. It has high resistance to mildew and is not attacked by moths.

Acele, Carolan, Celacrimp, Celafibre, Dicel, Estrella and Crycloset are some of the trade names of acetate fibre. Some of the chief properties of triacetate fibre, Tricel, are given below:

a. It is similar in strength and appearance to acetate but is more robust in wash and wear.
b. It has heat-setting properties and suitably constructed fabrics can be permanently pleated.
c. It has excellent resistance to shrinkage and stretching.
d. It resists soiling and is easy to wash.
e. It is quick drying and suitably constructed and finished fabrics made from this fibre require little or no ironing.
f. It has an exceptionally high resistance to degradation under hot wet conditions such as encountered in laundries.

Arnel, Tricel, Saflon, etc., are some of the trade names of triacetate fibres.

4.7.5 Bamboo fibre

Bamboo plant grows more rapidly than trees, and start to field within 2–5 years of planting. Bamboo can be selectively harvested and non-distinctively. There are more than 200 species of bamboo recognized. The establishment of bamboo plantation requires minimal capital investment and builds upon the plant cultivation skills of local formers and foresters. Bamboo is excellent for storing degraded lands and protecting against soil erosion. Bamboo may be easily intercropped with vegetables. The whole bamboo plant is beneficial for rural livelihood. The poles (technically referred as to 'culms') are useful as construction material; the young shoots are edible; leaves make good animal fodder; and branches are useful for making handicrafts besides its usefulness as clothing fibre.

Bamboo fibre is a kind of regenerated fibre, which is produced from raw materials of bamboo pulp. Firstly, bamboo pulp is refined from bamboo through a process of hydrolysis alkalization and multi-phase bleaching. Pulp is then processed in the normal root to produce fibre.

The type of bamboo is used for making fabric is commonly known as "Meso". It can reach to a height of 75 ft in just 45–60 days. Bamboo not only grows quickly but rarely needs pesticides. If there is sufficient rainfall, no additional irrigation is required. It regenerates naturally through an extensive root system that sends out an average of four to six new shoots per year.

Bamboo fibre products posses 3 times more effective antibacterial property, 12 times more effective antistatic property, 60% more effective improvement in hygroscopic properties and 30% more effective improvement in deodorization properties as compared with cotton products.

Table 4.10 Physical properties of bamboo fibres

	Properties	**Data**
1	Dry tensile strength (cN/dtex)	2.33
2	Wet tensile strength (cN/dtex)	1.37
3	Dry elongation at break (%)	23.8
4	Linear density (% deviation)	1.8
5	Percentage length deviation	1.8
6	Over length staple fibres (%)	0.2
7	Whiteness (%)	69.6
8	Moisture region	13.03
9	Oil content	0.17

(*Note*: Test done at 20 °C and 65% RH).

Bamboo fibre is ecofriendly and biodegradable and has unique antibacterial, moisture absorption, deodorant, UV resistant, colouration, elasticity, drapability and wearability characteristics. Because of its special structure and natural hollowness in the horizontal cross sections, the abundant gaps is the fibre can absorb and evaporate the moisture of human skin instantly, thus gives soft feel. Moisture absorbency is twice than that of cotton with extraordinary soil release.

Bamboo fibre has natural antibacterial elements (bio agent "bamboo Kun") and show great elasticity. It is one of the strongest building materials. Bamboo fibre finds its use for knitted underwear, T-shirts, shirts, mats, socks, hygiene products, bath suits, etc.

4.7.6 Soyabean protein fibre

Soyabean fibre is a protein fibre. It is processed similar to any other proteins, viz. casein fibre from milk, saran fibre from maze, etc. depending upon the chemical composition of the raw material. These protein fibres are in generally have warmth feeling properties.

Breaking strength of single soyabean fibre is over 3 cN/dtex which is higher than that of wool, cotton and silk. By now 1.27 dtex fibre can be spun into 6 dtex yarn which can be used for high quality and high density fabrics. Since initial modulus of soyabean fibre is on higher side, boiling water shrinkage is found to be low. Soyabean fibre has lustre of silk and possess excellent drape. The cross-section of soyabean fibre is 'dumb bell' shape, vertical shear of fibre has longitudinal grooves.

Table 4.11 Comparison of soyabean protein fibres (SPF) with other fibres

Property	**SPF**	**Cotton**	**Viscose**	**Silk**	**Wool**
Breaking strength (cN/dtex) (a) Dry (b) Wet	3.8–4.0 2.5–3.0	1.9–3.1 2.3–3.1	1.5–2.0 0.7–1.1	2.6–3.5 1.9–2.5	0.9–1.6 0.7–1.3
Dry breaking strength (%)	18–21	7–10	18–24	14–25	25–35
Initial modulus (kg/mm2)	700–1300	850–1200	850–1150	650–1250	–
Moisture region (%) Density (g/cm3)	8.6 1.29	9.0 1.5–1.54	13.0 1.46–1.52	11.0 1.34–1.38	14.16 1.33
Heat endurance	Yellowing at 120 °C (bad)	Brown after long time at 150 °C (excellent)	Strength down after long time processing at 150 °C (good)	Stable at temperature ≤148 °C (good)	Good
Alkali resistance	General level	Excellent	Excellent	Good	Bad

Acid resistance	Excellent	Bad	Bad	Excellent	Excellent
UV	Good	General level	Bad	Bad	Bad

The soyabean fibre contains lots of pofar amino acids necessary for the human body. The antibacterial agents those are added to the soyabean fibre in spinning process can restrain the growth of colon bacillus, impetigo bacterial and sporothrix. Therefore, soyabean fibre is a kind of 'sanitarian' fibre.

4.8 Synthetic fibres

Synthetic fibres are those in which man performs the entire operation of fibre manufacture without allowing nature to intervene. The chemicals used are made into materials capable of forming fibres and these materials are then manipulated into fibrous form. Therefore, synthetic fibres are the fibres manufactured by synthesis of chemicals.

All synthetic fibres now being produced have a number of characteristics in common. They are based on fibre forming polymer (poly = many, mer = unit) with some common properties, they are made by similar process from these polymers. Polymer is a union of simple monomers (mono = one), into a giant molecule called macromolecule. The process of joining the monomers is called polymerization and the length of such chains is referred to as the degree of polymerization (DP). In other words, if a substance which has a reactive or functional group at either end of its molecules, then two of its molecules could combine together to yield a molecule twice as long, which also has a reactive or functional group at both the ends resulting in the repetition of the process. This repeated process of doubling or synthesis of molecules is known as polymerization. Therefore, polymerization is a process whereby one compound is transformed into another having the same elements in the same proportion but different molecular weights and different physical properties.

There are mainly three types of polymerization, viz.

a. Condensation type

b. Addition type

c. Substituted type

If linkage of monomer molecules takes place by chemical action which results in the elimination of by-product molecule like water, hydrogen chloride or ammonia, it is called condensation polymerization. Polyamide and polyester are the examples of fibre forming polymer made by condensation polymerization.

If the linkage of monomer molecule comprises simply the union of several molecules but with no splitting off of water or hydrogen chloride, it is called addition polymerization. Polyethylene, polypropylene, polyvinyl chloride are examples of fibre forming polymer made by addition polymerization.

The notable difference between the two methods of polymerization-condensation and addition type is that, in the former simple compound is split off, and in the later nothing is split off. As the name indicates, addition polymerization consists of addition of one molecule to another. This is brought about by the application of high pressure and high temperatures usually in the presence of a catalyst.

In the substituted type, the linkage of monomer molecule is simply substituting for another group. For example, halogen substituted type polyvinylidene chloride, saran. If it is substituted by hydroxyl group it yields fibre like polyvinyl alcohol (vinal). If it is substituted by nitrile group, it yields polyacrylonitrile and vinylidelle dinitrile, etc.

Substances that result in polymerization reaction are known as linear polymers because they are long, straight chain compounds. This has provided fibre-scientists an approach to a method of making fibrous molecules which in bulk would constitute fibre forming material. The solid fibre forming material is rendered into fluid state. This is done either by (a) dissolving the material in a solvent or (b) by heating it until it melts. Subsequently it is spun into yarns either by solvent spinning or melt spinning.

In recent years, much public interest has been shown in the so-called synthetic fibres, and the first of these, termed nylon, was introduced in 1937 by E. I. Dupont de Nemours Co. Inc. in America.

The credit of inventing first synthetic fibre goes to Dr. W.H. Carothers; from then onwards lots of fibres have been added to the list of synthetic fibres. The synthetic fibres entered Indian scene in the early 1960. In the production of total manmade fibres in India during the period 1966–1979, the contribution of synthetics initially was very low, it rose to 8%, 12% and 25% of the total production during the years 1968, 1971 and 1980, respectively.

Global scenario

Individual fibre production patterns have resulted in significant shifts in fibre shares. Polyester production share has moved from 37% in 1982 to 58% in 2002, a 21% absolute share gain. Olefin share has grown from 7% in 1982 to 17% in 2002. From 1982–2002 trend growth for all manufactured fibres was 4.8%. For the 20 years period, synthetics grew at 5.7% per year, while cellulosics declined at a rate of 1.6% per year. Within the synthetic group, olefin grew at 9.0%, polyester at 7.3%, acrylic at 1.4% and nylon at 1.6%.

Global production of manufactured fibre was 36.0 million metric tons in 2002, an increase of 155% from the 14.1 million tons produced in 1982. Over the past 20 years, synthetic types, e.g. polyester, have shown strong long-term growth, while celluloses have declined. Synthetics now account for 94% of worldwide production. Cellulosic fibre production peaked in the early 1980s at 3 million metric tons. Since then, the regional trend has been characterized by strong growth in Asia offset by drastic decline in Eastern Europe – from peak production of 1.1 million metric tons in the early 1980s to 92,000 tons in 2002. In the same period Asian production moved up by 660,000 tons and accounted for 69% of global production in 2002.

Asian share of synthetic fibre production was 65% in 2002. This compares with a North American and Western Europe combined share of 46% in 1982, down to 27% in 2002. The other significant shift in the 20 years period was Eastern Europe – moving from 17% to 2%. Other regions in that period including South America, the Middle East and Oceania experienced consistent growth. Table 4.12 gives the figures of world synthetic fibre production from 1900 to 2013

Table 4.12 World manmade fibre production

	Raw cotton (5)	**Raw wool (3) (6)**	**Manmade fibres (1)**							**Total**
			Symbolics (2)			**Cellulosics**			**Total**	
			Yarn	**Staple**	**Total**	**Yarn**	**Staple (4)**	**Total**		
1900	3162	730	.	.	.	1	.	1	1	3893
1910	4200	803	–	–	–	5	–	5	5	5003
1920	4629	816	–	–	–	15	–	15	15	5460
1930	5870	1002	–	–	–	205	3	208	208	7030
1940	6907	1134	1	4	5	542	585	1127	1132	9173
1950	6647	1057	54	15	69	871	737	1608	1677	9381
1960	10113	1463	417	235	702	1131	1525	2656	3358	14934
1980	13844	1599	4743	5882	10625		3557	3557	14132	29625
1985	17383	1744	5811	7307	13118		3218	3218	16336	35463
1986	15339	1789	6046	7630	13676		3178	3178	16854	33982
1987	17670	1832	6378	8098	14476		3236	3236	17712	37214
1988	18366	1886	6947	8285	15232		3284	3284	18516	38768
1989	17431	1955	7246	8472	15718		3284	3284	19002	38388

1990	18997	1927	7173	8201	15374		3145	3145	18519	39443
1991	20793	1779	7369	8358	15727		2897	2897	18624	41196
1992	17990	1719	7786	8549	16335	692	2045	2737	19072	38781
1993	16673	1673	8016	8506	16522	670	2033	2752	19274	37620
1994	18695	1554	9293	9152	18445	672	2204	2677	21322	41571
1995	19962	1486	10169	9021	19190	667	2347	3014	22204	43652
1996	18960	1476	12180	10285	22466	649	2211	2859	25325	45761
1997	19849	1440	13778	11241	25020	625	2271	2396	27915	49204
1998	18429	1382	15084	11505	26589	591	2142	2733	29322	49133
1999	19176	1380	16203	12011	28214	520	2117	2637	30351	51407
2000	18901	1357	17551	12737	30288	498	2297	2795	33083	53341
2001	21237	1309	17719	12596	30315	483	2201	2684	32999	55545
2002	19142	1254	19031	13213	32244	466	2276	2742	34986	55382
2003	20430	1233	20016	13753	33769	485	2399	2383	36652	58315
2004	26126	1221	21403	14705	36103	498	2654	3152	39260	66607
2005	24763	1228	22706	15431	38137	455	2699	3154	41291	67282
2006	26452	1229	24155	15752	39907	448	2858	3306	43213	70894
2007	26150	1200	26562	16596	43158	447	3083	3535	46693	74043
2003	23330	1100	25750	15331	41081	398	2836	3234	44315	68745
2009	22000	10S0	26551	15964	42515	359	3530	3889	46404	69484
2010	24450	1062	30509	16887	47396	377	3686	4063	51459	76971
2011	26925	1068	33106	17662	50768	387	4113	4500	55268	83261
2012	26785	1073	36133	13522	54655	401	4783	5184	59839	87697
2013	25725	1090	38665	18810	57475	375	5336	5711	63186	90001

(Courtesy: CIRFS@ European Manmade Fibre Association).

In 2002 global production – in million metric tons – for the key fibre types was: polyester 21.0, olefin 5.9, nylon 3.9 and acrylic 2.7. Over the past 20 years, polyester has maintained a substantial lead in volume. Olefin fibres have shown strong growth from a low base and now surpass nylon and acrylic fibres in production volume.

Growth in synthetic fibre production has resulted in a major shift in production from North America and Europe to Asia. In 2002, Asian production was 20.0 million metric tons compared with 4.7 in North America, 3.4 in Europe. Production in Eastern Europe has declined significantly – from a peak of 1.9 million metric tons in the late 1980s to 0.9 in 2002.

The 20 year average annual growth trend by region reflects 8.8% growth in Asia, 2.1% growth in North America, 1.9% in Western Europe and 4.3% for all other regions. Eastern Europe declined at 8.7% per year from a peak production in 1988.

In 1997 global growth was 11.5%, strikingly above trend line growth of 5% per year. In 1998 growth declined to 3.4%, precipitated by the Asian currency crisis but aggravated by over built global inventories and several years of over expansion of global capacity forcing production to unsustainable growth rates. Growth accelerated in 1999 and 2000 to 4.6% and 6.6%, respectively. In 2001, for the first time in history, worldwide synthetic production declined by 0.6% to be followed by a 6.1% increase in 2002 (see Fig. 4.12).

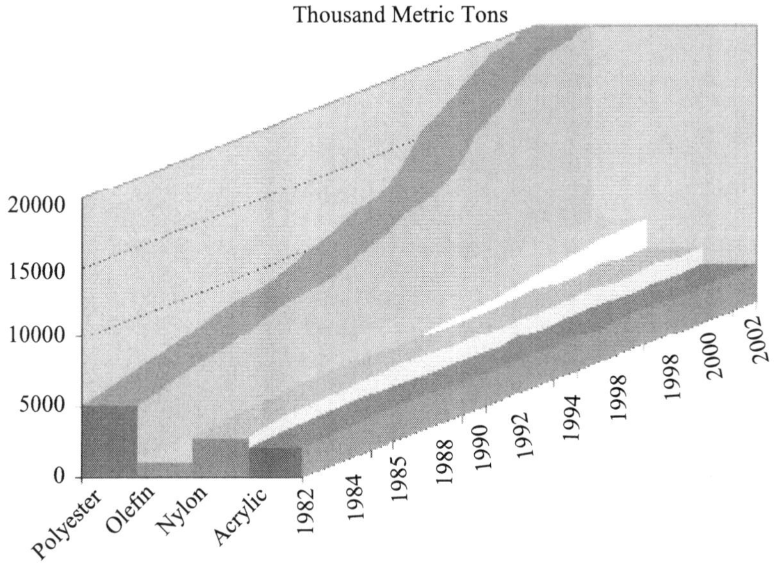

Figure 4.12 Worldwide synthetic fibre production by fibre type (thousand metric tons).

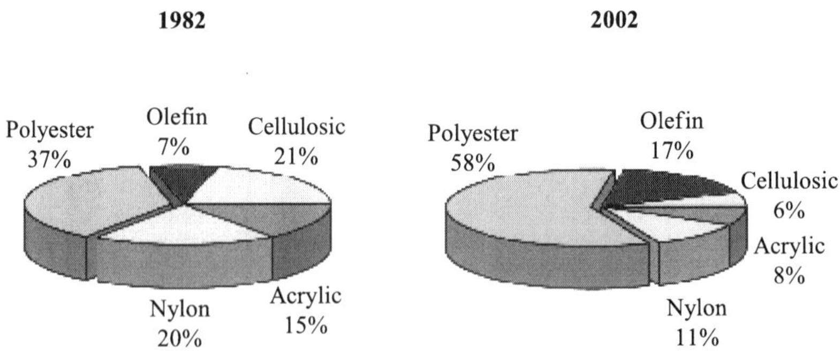

Figure 4.13 Shift in world synthetic fibre production 1982 vs 2002.

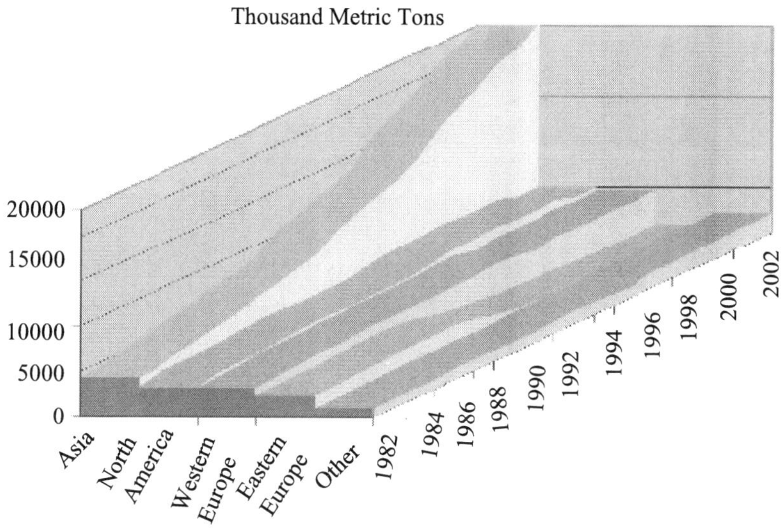

Source: The above data (Figs. 4.12–4.14) is drawn from annual World Surveys of Manufactured Fiber Production, completed each year by the Fiber Economics Bureau and published in June issue of the Bureau's monthly statistical journal, The Fiber Organon.

Figure 4.14 Worldwide synthetic fibre production by region (thousand metric tons).

4.8.1 Manufacture of polyamide fibres

Nylon has been accepted as a generic or collective term representing a family of fibre forming polyamides. The most important types are Nylon 6 and Nylon 66.

Nylon 6 is made from caprolactam, which is made by a series of reactions using the products obtained from coal tar. The main reactions in the production of caprolactam are shown in Chart 4.2.

Caprolactam is a white flaky solid, melts at 68 °C and is soluble in water. The polymerization is carried out in stainless steel cylinders of 1.8–2.7 m in height and 0.5 m in diameter. It is maintained at 240–270 °C temperature by vapour-heated or electrically heated jackets. Hot caprolactam is mixed with a suspension of pigment. A solution of an acid promoter as well as an acid chain-stopper is fed continuously at the top of the cylinder. During the downward flow, polymerization takes place and at the bottom of the cylinder a high viscosity molten polymer, consisting of 89–92% of the polymer and 8–11% of the unconverted caprolactam.

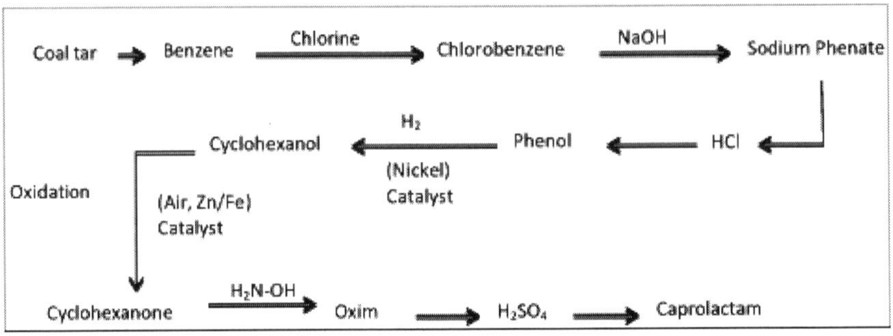

Chart 4.2 Main chemical reactions in the production of caprolactam.

A detailed process sequence of the manufacture of nylon-6 fibre is given in the flow sheet chart. It may be consulted along with the flow diagram given in Fig. 4.6 for better understanding of the process.

The extent of conversion of caprolactam to the polymer depends upon the temperature of the polymerization. For example, 89% conversion can be achieved at 260 °C. The acid chain-stopper may be acetic acid, stearic acid or adipic acid, which at a particular stage of polymerization reacts with the amino group, present at one end of the macromolecule and prevents further polymerization. Hence, a desired viscosity of the molten polymer may be obtained.

The molten material collected at the bottom of the polymerizing unit is handled in a separate unit under reduced pressure when the relatively volatile caprolactam (unreacted) is removed by evaporation. The molten polymer is extruded into ribbons, solidified by water and cut into chips.

The chips are used for the production of continuous filaments. The molten polymer may be directly used (without preparing the chips) for the production of staple fibres. For making continuous filaments, the chips are washed with water at 80 °C and caprolactam, still remaining in the polymer is removed by centrifuge. It is further centrifuged to remove the excess water and then dried under reduced pressure in tumble driers

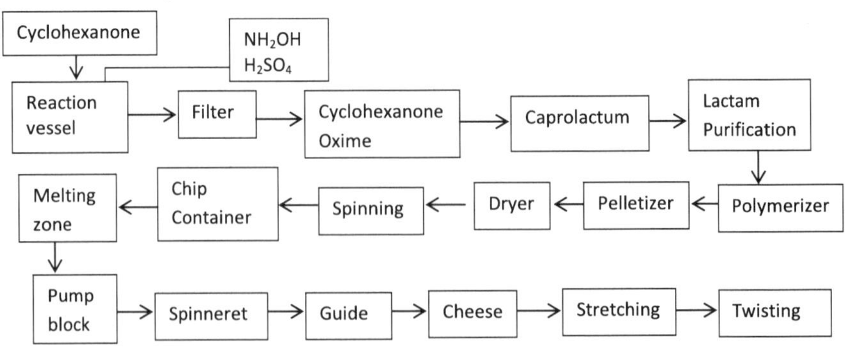

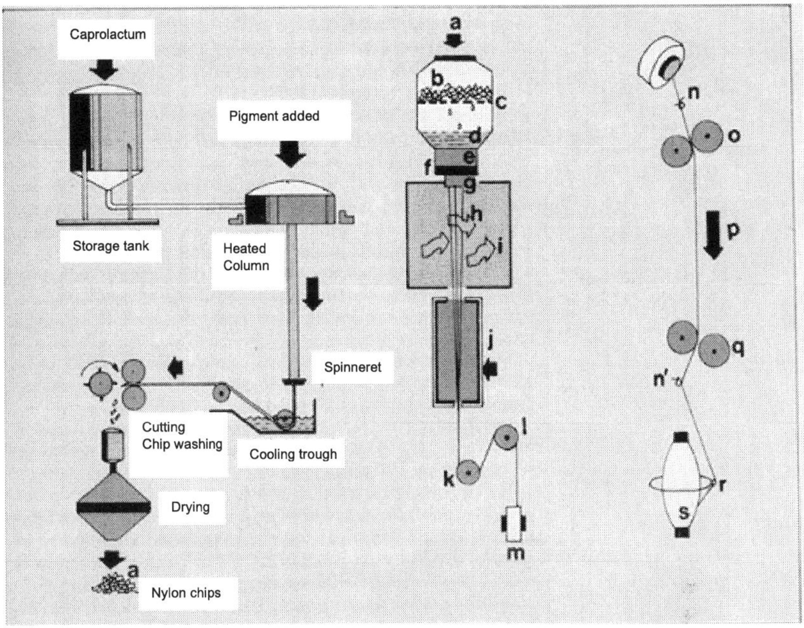

Figure 4.15 Flow diagram of nylon 6 manufacture.

Continuous filaments are made by using melt spinning technique (Fig. 4.15). Dry polymer chips (a) are fed to a melt spinning apparatus (b), wherein one section of the chips falls into a melting region (c), where they are heated electrically to 250–260 °C. The molten polymer flows into a conical section to form a pool (d), which feeds a spinning pump (e) and spinneret (h). The pool is kept under an atmosphere of nitrogen to prevent decomposition by air. The whole spinning apparatus is made of stainless steel and is properly insulated to prevent loss of heat.

The molten polymer leaving pump is filtered by passing through a layer of sand packs of graded gauge filters (f), before entering the spinneret, which is a stainless steel disc, having a number of fine holes. The number and diameter of which determines the type of yarn formed. The molten polymer coming out of the spinneret (g), i.e. the spinning jet (h) solidified in falling through the air (i), to a take-up machine (j) which is about 6 m below the spinneret.

Before reaching the mandrel on which the cheese is built up, the filaments are moistened by passing them through a steam chamber (j) to ensure dimensional stability of the final package which may contain about a kilogram of filament.

The take-up machine operates at a very high speed of about 1000 m/min. About 45–50% RH and 20 °C temperature is maintained around take-up machine.

The filament thus obtained will not be strong enough and has a very high extensibility. It has to be subjected to a treatment of stretching to produce the desired textile properties.

The mandrel holding the unoriented yarn is placed at the top of the stretching unit, and the filament yam is taken through the stretching or drawing device, which consists of an upper set of nip rollers (o), running at 100 m/min, and a lower set of rollers (q) running at 300–400 m/min, and hence stretch the yarn 3–4 times its original length. Drawing is carried out at room temperature (cold drawing). A deflector (p) kept in between the top and bottom drawing roller (o and q) control the movement of the filaments in the drawing zone. Thread guide (n) is provided at strategic point to guide the filament properly to the required position. Finally, the filament yarn which is delivered from the lower nip rollers is twisted on uptwister (r) operating at a speed of 8000 rpm, and, is collected on the spindles.

The twist per metre on the yarn depends on the end-use. For example, hosiery yarn is imparted a twist of about 1182 tpm (30 tpi) while for weaving purpose a low twist of about 275.8 tpm (7 tpi) is considered sufficient.

The residual caprolactam (about 3%) on the filament yarn is removed by washing with boiling or nearly boiling water, which also sets the twist in the yarn.

Nylon filaments can be obtained in the multifilament or monofilament form.

Multifilament is made up of a number of tiny, almost endless strands twisted together into one yarn. The size and the number of these minute strands can be varied as well as the amount of twist. Most of the nylon fabrics are made from this versatile type of filament. Multifilament yarn gives a pleasant surface texture, softness and luxurious drape.

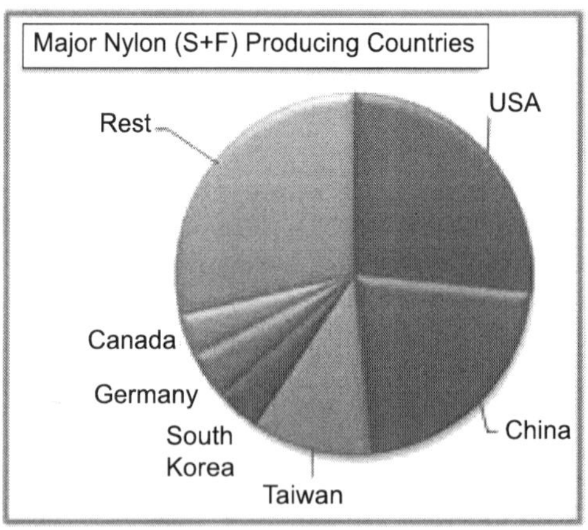

Figure 4.16 Major nylon producing countries.

Source: www.yarnsandfibers.com for the year 2008.

Monofilament is single, solid strand of filament of great strength and smoothness. Approximately 90% of all polyamide fibres are processed in the form of monofilaments. Its main end-uses are car tyres, fishing ropes and nets, gliders tow ropes, sail cloths, conveyor belts, carpets, upholstery, draperies, lingerie, women's sheer stocking, etc.

4.8.2 Some similarities and differences between nylon-6 and nylon-66 fibres

The suffix numbers to the word nylon relate to the number of carbon atoms present in the basic compounds, which form the polymer. Nylon 6 is made from one component, namely, caprolactam, which has six carbon atoms, while nylon 66 is made from two components, namely, adipic acid and hexamethylenediamine, each of which has six carbon atoms. These basic substances are made by complicated chemical processes from coal tar, or petrochemical products.

Nylon 6 is very similar in general properties to nylon 66. It is claimed that it is softer in handle than nylon 66, and that it stays white and does not turn yellow in the way that nylon 66 sometimes does.

Both nylon 6 and nylon 66 are used for clothing purposes. Nylon 6 has lower melting point (218 °C) than nylon 66 (250 °C). Nylon 6 has greater dye affinity for certain dyestuffs than nylon 66. Even in spite of its low melting point nylon 6 has better heat stability.

Both nylon 6 and nylon 66 are sensitive to ultraviolet light and tend to undergo degradation. Resistance of nylon 6 is far greater than that of nylon 66.

Table 4.13 Comparison of the properties of nylon 6 and nylon 66

Properties	Nylon 6	Nylon 66
Tenacity (gpd)	6.0–9.5	3.0–6.0
Elongation (%)	23	23
Moisture regain (%)	4.5	4–4.5
Melting point (°C)	218	250
Sticking point (°C)	171	229
Heat setting temperature (°C)	150	205
Density	1.14	1.14
Safe ironing temperature (°C)	149	177

Nylon 6 possesses better elastic recovery and fatigue resistance then nylon 66. A comparison of other properties of nylon 6 and 66 is given in Table 4.13.

4.8.3 Manufacture of polyester fibre

This class of fibres was discovered in 1941 by J.R. Whinfield and J.J. Dickson et al. of UK put to commercial production in the early 1950s. These fibres are manufactured under different names such as Recron, Dacron, etc.

The polyester (PET) is a condensation polymer and is made by heating terephthalic acid or dimethyl terephthalate with excess of ethylene glycol in an atmosphere of nitrogen initially at atmospheric pressure. A catalyst is used to speed up the reaction. The excess of ethylene glycol is distilled off with steam at a pressure of 450 psi, in ester exchange vessel. The ester can polymerize now to form a product of high molecular weight. The resulting low molecular weight ethylene glycol terephthalate is then heated at 280 °C, for 30 min at atmospheric pressure and then for 10 h under vacuum.

The polymer obtained in the above manner is hard and almost white. It is extruded in the form of ribbon from the autoclave, which is cut into chips by the cutter. This is followed by the drying (at 120 °C) operation to remove the residual moisture.

The chips are discharged from the hopper reservoirs, and are ready for melt spinning.

Polyethylene terephthalate melts at 260 °C and the molten mass is stable as long as oxygen is effectively excluded. Every care should be taken to avoid the air coming into contact with molten mass.

Spinning of filament is carried out in a manner similar to that of polyamide, but the main advantage is that the individual filament solidifies instantaneously. A spin finish is given at that stage with a lubricant and an antistatic agent. The undrawn yarn is then wound onto a cylinder at a speed of 1000 m/min.

This filament yarn is taken to the drawing, where the undrawn filaments are drawn to about 4–5 times its original length. The stretching is carried out at elevated temperatures. Two varieties of polyester filaments are produced, viz. (1) normal strength yarns used mainly for clothing, and (2) high tenacity (HT) filaments are made stronger during processing and are used for industrial purposes.

Very coarse filaments owing to their poor thermal conductivity are cold drawn. The cold drawing is carried out at room temperature.

For the production of polyester staple fibre, extrusion of filament takes place on a much larger scale than filament yarns. The filaments are first brought together to form a thick tow. The tow is drawn to get the correct strength. The tow next passes through a crimping machine. The crimps are stabilized by heating in ovens. It is then cut into specified lengths and baled ready for dispatch. The flow sheet for the manufacture of polyester fibre, both for filament and staple, is given in Fig. 4.17

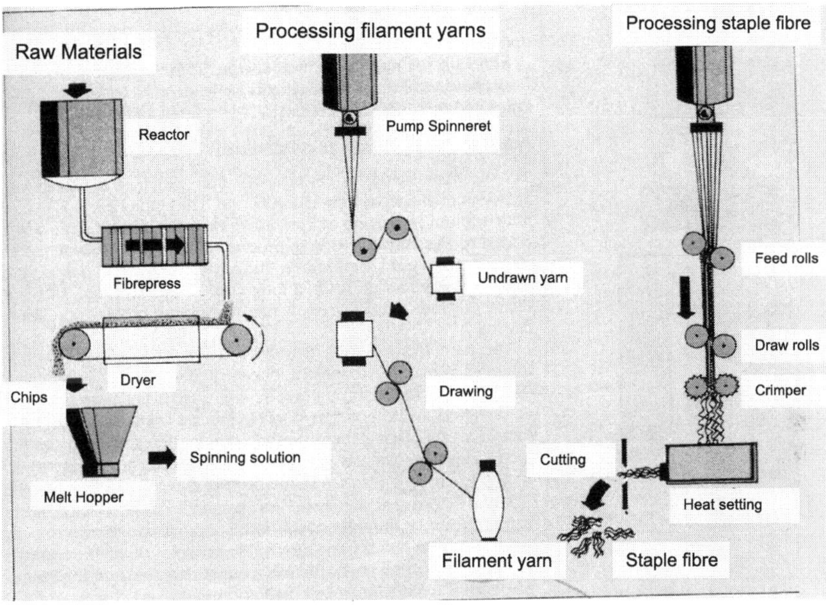

Figure 4.17 Flow diagram for the manufacture of polyester.

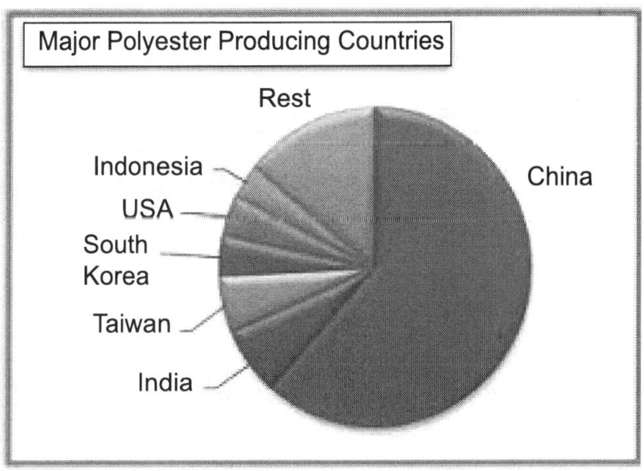

Figure 4.18 Major polyester producing countries.

Source: www.yarnsandfibers.com for the year 2010.

4.8.4 Manufacture of acrylic fibre

The youngest of three main synthetic fibres is polyacrylonitrile (Acrylic), which has made very rapid progress during the last two to three decades.

Acrylic is the generic name given to a closely related family of fibres, all having a chemical base of acrylonitrile, plus a minor constituent with which it has been co-polymerized to increase the dye affinity of the fibre. It is primarily this minor constituent which gives the individuality to the many acrylic fibres, all bearing different trade names. At least 30 different acrylic fibres are in existence now. Many of them are already well-known, such as, Acrilan, Courtelle, Creslan, Dralon, Zefran, Verel, Crylor, etc.

Acrylic fibres can be manufactured by two different methods. In the first method, the polymer may be dissolved in a suitable solvent such as dimethylformamide, extruded through spinneret, and the solvent evaporated off, this is known as dry spinning. In the second method, polymer solution is extruded into a heated precipitating bath, containing glycerol or any other suitable coagulating agent. This method approximates to the wet-spinning.

Although the machinery used is completely different in design, in both these techniques the filaments are stretched to orient the molecules and then improve the physical properties of the fibre.

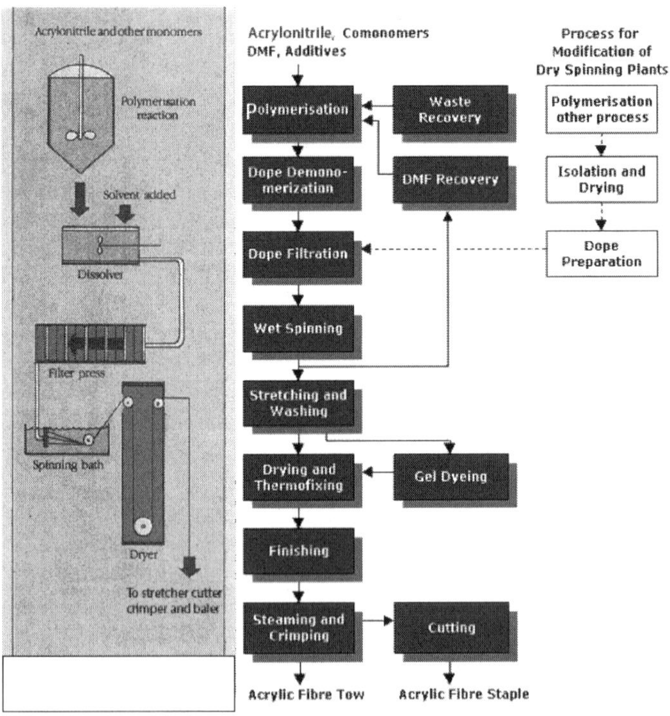

Figure 4.19 Flow diagram showing manufacture of an acrylic fibre.

For the manufacture of Orlon-acrylic fibre, ethylene is treated with hypochlorous acid to form chlorohydrin. The chlorohydrin is reacted with sodium hydroxide to form ethylene oxide. Hydrocyanic acid is added to the ethylene oxide to produce cyanoalcohol, which is then dehydrated to yield acrylonitrile. The acrylonitrile is then polymerized into polyacrylonitrile resin, a long chain linear polymer. This long chain polymer is converted to bits resembling broken noodles by cutting.

The hard bits are then dissolved with a suitable solvent, such as dimethylformamide, to make a concentrated solution. After filtration, the solution is forced through a spinneret, dried in hot gases and drawn or stretched, while hot. The drawing process orients the long chain molecules by placing them parallel, running lengthwise in the filament. Drawing controls the stretch of the fibre and improves its strength. The flow diagram of the process is represented in Fig. 4.19.

About 98% of polyacrylics are manufactured as staple fibres. Yarns from acrylic fibres have high bulk, and fabrics woven from it give a warm wool-like handle. The high bulking process of acrylics is used in the fashion knitting trade. Blankets, carpets and outerwear garments made from acrylics are lighter than those produced from wool. Of all the manmade fibres acrylics have the least tendency to pilling.

4.8.5 Modacrylic fibres

Modacrylic fibres are based on acrylonitrile and vinylidene chloride, derivatives of oil refining and coal carbonization processes. Modacrylics are manufactured fibres in which the fibre-forming substance is any long chain of synthetic polymer composed of less than 85%, but at least 35% by weight, of acrylonitrile units.

- Modacrylic fibres are produced under different trade names such as, Dynel Teklan, Verel, etc.
- Modacrylic fibres have 55% elastic recovery at 10% stretch, 88% at 4% extension, while acrylics possess 50% elastic recovery at 10% stretch and 90–95% are 1% extension.
- Modacrylic fibres (Verel) are produced in peanut-shaped cross-section while wet spun acrylics have a round or kidney shaped cross-section and dry spun fibre has dog bone cross-section.
- Modacrylic fibres have initial modulus of 28 gpd, compared to acrylic fibre which has high initial modulus value of 40–50 gpd.
- Modacrylic fibres stick at 200–204 °C whereas acrylic fibres have higher sticking temperature of 215–255 °C.

- Modacrylic fibres in addition to their aesthetic, silk-like qualities are strong, hard-wearing and inherently flame resistant.

Uses of modacrylic fibres include, woven and knitted dress-goods, lingeries including brushed materials; overall fabrics; pile fabrics and knitwear in apparel textiles, curtain nets; curtain drapes; theatre curtains; etc., in household textiles, filter cloths, electrical insulation, flame resistant reinforcements in industrial uses.

4.8.6 Bi-component fibres

Bi-component fibres are those fibres which may consist of two or more components of different polymers and extruded as a single filament.

Bi-component fibres consist of two components divided along the length of the fibre into two more or less distinct regions.

Bi-component fibres are also referred to sometimes as composite, conjugate, and hetera fibres. Some different types of bi-component fibre cross-sections are illustrated in Fig. 4.20.

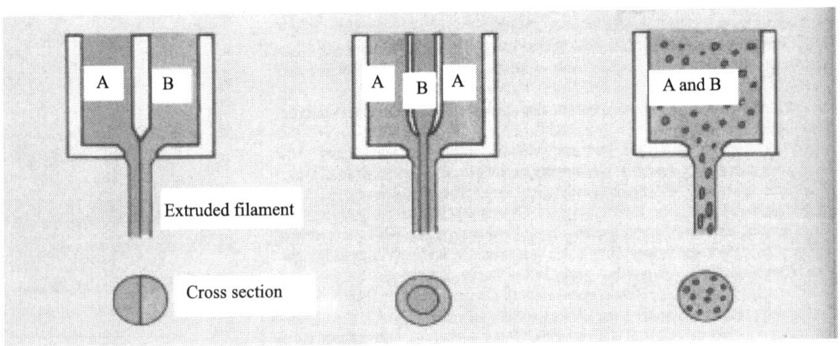

Figure 4.20 The principles of bi-component spinning and the cross-section of fibres obtained.

Manufacturing of side-by-side component fibres can be classified into three groups:

- In the first group, the two components either solutions or melts are fed directly to the spinneret orifices, being combined into bi-component fibres at or near the orifices.
- In the second group, the two bi-components of the fibre are formed into a multi-layered structure (either flat sheets or concentric cylinders) and fed without turbulence to the spinneret, the rows of orifices in

the spinneret being so positioned as to intersect the interfaces of the various layers of polymer.

- In the third group, the two components are also formed into a non-turbulent layer structure (e.g. a mixed stream) and fed to the spinnerets; here, however, no attempt is made to align the rows of spinneret orifices with the component interfaces, and so bi-component fibres of wide range of compositions are produced

It is possible that a bi-component fibre may vary in the lateral distribution of the two components along the fibre length. It is also possible to prepare yarns consisting of mixture of bi-component fibres and monocomponent fibres.

These bi-component fibres are suitable for the production of bonded fabrics, floor coverings, upholstery fabrics, high crease resistant fabrics, etc.

Use of bi-component fibre in manufacturing bulked yarn for knitting: Acrylic bi-component fibres exhibit a reversible crimp on steam wetting and drying, similar to what is observed in the case of wool. In fact, acrylic bi-component fibres were developed after studying the morphological structure of wool fibre. The typical hand and bulkiness of wool fibres is largely attributed to the bilateral structural asymmetry of the fibre which is due to the presence of two components namely ortho and para-cortex units in the cross-section. Acrylic bi-component fibre is normally prepared by spinning fibres from the acrylic copolymers having different longitudinal shrinkage characteristics.

When bi-component fibre is subjected to heat, differential shrinkage is developed in the fibre itself and as a result a three-dimensional crimp is formed. This three-dimensional crimp in the fibre in turn generates a very good feel, high elasticity, high resiliency and improved dimensional stability of the yarn. The crimp configuration of acryli bi-component as well as monocomponent fibre in producing bulky hand knitting yarns is as shown in Fig. 4.21

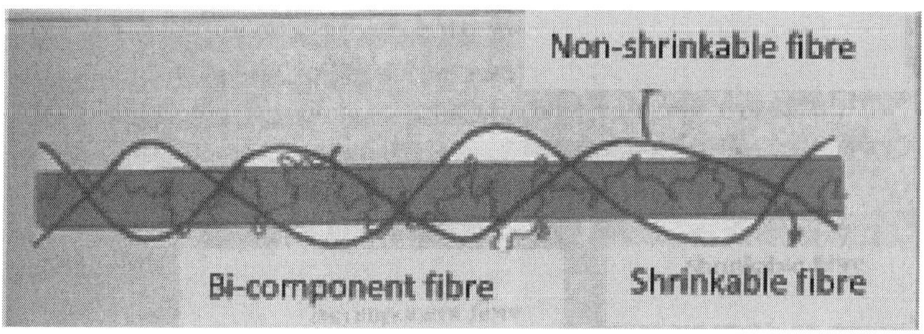

Figure 4.21 Bulky acrylic yarn using shrinkable, non-shrinkable and bi-component fibres.

4.9 Inorganic fibres

4.9.1 Manufacture of glass fibre

The idea of making yarns and fabrics from glass is centuries old. The raw materials for glass are primarily silica sand and limestone, with small amount of other compounds such as aluminium hydroxide, sodium carbonate and borax. The formulation of the glass depends upon the end-use. Games Slayter, John H. Thomas and several other associates from America, developed a process of drawing fibrous glass in 1931. Within a year they introduced coarse glass fibres formed into a pack and treated with a dust stopping adhesive for use as air filter. Until 1936, the glass fibres that were developed were not suitable for textile purposes. In 1937, fibres were prepared commercially which were pliable enough to be woven into all-glass fabrics.

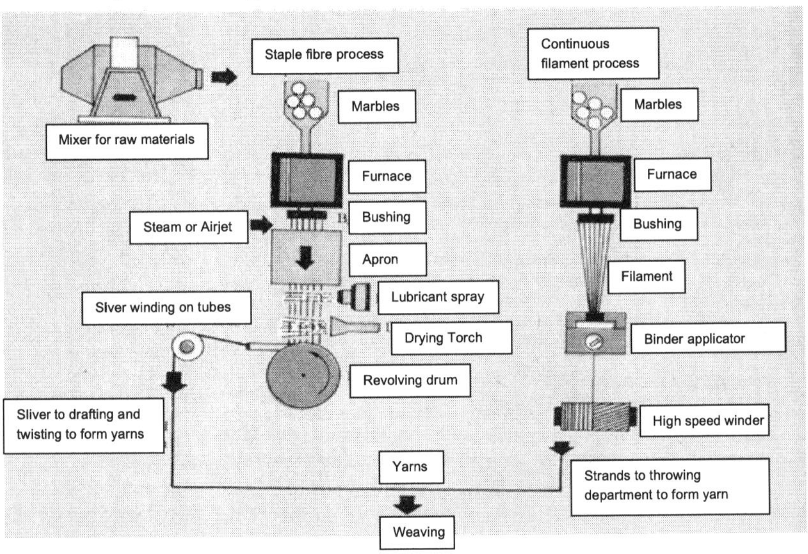

Figure 4.22 Flow diagram showing glass fibre manufacture.

Following selection of the raw materials depending upon the ultimate end-use, they are melted in a furnace at temperature of about 3000 °F. The compounds combine to form a clear glass marble of about 1.6 cm in diameter. The marbles are called gullets. Imperfect gullets are sorted out to prevent imperfections in the filaments.The perfect marbles are fed into a small furnace, where they are melted at temperature of about 2500 °F (1371 °C) and the molten glass falls by gravity pull through a platinum spinneret which has approximately 100 orifices. As the molten glass leaves the spinneret, it solidifies. Flow diagram of glass fibre manufacturing, of both staple and filament process, is given in Fig. 4.22. For filament yarns the fibres are pulled together, lubricated for ease in handling, and wound on tubes in strand form, for use in tabric manufacture. Each marble of 0.0284 kg can produce about 62 km of filament

4.9.2 Metalic: fibres or threads

Metallic fibres are fibres composed of metal, plastic coated metal, metal coated plastic or a core completely covered by metal. These are light weight and relatively inexpensive yarns.

Any of the ductile metals such as gold, silver, copper or even the cheaper alloys, can be drawn out into fine filaments. Metal threads are usually made with a core of cotton yarn wrapped within a ribbon of metal. Gold thread and silver wire are used for expensive fabrics. Imitation gold and silver wires are also prevalent, but they get tarnished easily.

Modern metallic yarns are produced from coloured aluminium ribbons sandwiched between two layers of transparent plastic material. Aluminium, the basic metal used is softer, lighter in weight and cheaper than the more precious metals and is not apt to tarnish or cause discolouration.

The use of metallic yarns is affected tremendously by the dictates of current fashion. There are very few items in which metallic threads have not appeared.

5
Fibre structure

The structure of a fibre is described in terms of degree of polymerization, molecular orientation, molecular bonds, amorphous region crystallinity and crosslinks. Sophisticated techniques like electron microscope, X-ray and birefringence studies, etc., reveal the fibre structure. In this chapter fibre structure and related properties are presented.

5.1 Fine structure: General features

The natural fibres are made up of long, thin, complicated molecules of cellulose or protein as the case may be, and each of which contains thousands of atoms of a few elements such as carbon, hydrogen, oxygen, nitrogen, etc. For example, an average molecule of cotton contains as much as 40,000–80,000 atoms. Such molecules are called *macromolecules*. In most of the high molecular compounds the molecules are built up of a great number of similar elementary units and, therefore, are often referred to as high polymers or simply polymers. In general, a polymer is made by joining together of a large number of small molecules. It may be of the same type of molecule or of a mixture of different types of molecules, but seldom more than two or three types. These small molecules may join end-to-end to form straight linear polymer molecules or they may join end-to-end and also sideways, so as to form three-dimensional or branched polymer molecules. It is those polymers having simple long chain linear molecules which are most suitable for the manufacture of synthetic fibres.

The simplest method for forming a linear polymer is to select a substance (termed a monomeric substance or a monomer) whose small molecules have a tendency to join together and then hold this under appropriate conditions to induce the joining up process (termed polymerization) to continue until tests indicate that the polymer molecules produced have the desired length and other fibre-forming characteristics.

5.1.1 Arrangement of molecules

It is true in all fibre forming substances, natural and manmade that their molecules may be compared to a string of beads. The molecular chains lie along the length of the fibre. Just as fibres lie along the length of the yarn, so that the fibres may be regarded as molecular yarn. And just as the ability to form yarn depends on the length of the fibres, so also a certain minimum length of molecule is needed before a coherent fibre is formed. The greater the fibre length, the greater are the surfaces of contacts and the friction, and hence the yarn strength. Inside the fibre it is a kind of friction between the molecules that prevents them from slipping past one another, and so confers strength on fibre as a whole.

From the above explanation it is evident that there is a close and real analogy between the arrangement of the fibres in a yarn and the arrangement of fibrous molecules in a fibre.

The main sources of evidence of this observation are:

a. Anisotropic swelling

b. Birefringence

c. X-ray crystallinity

It is advisable to consider the analytical data of these a little more carefully.

Long molecules are essential for fibre strength. Fig. 5.1 illustrates how the area of attraction that has to be broken is greater in the case of long than of short molecules. We are thus able to see why fibres which consist of very long molecules will be stronger than those consisting of shorter molecules.

It is, however, important to note at this stage that once the chain length has reached that stage at which the energy required to make a molecule slide over its neighbours is greater than required to break a molecule, then further increase in molecular chain length will have no effect on the fibre strength. An idea frequently expressed is that the strength of a fibre increases with the length of its constituent molecules. This is correct, until a stage is reached when molecular slip is stopped, but beyond this it will be incorrect.

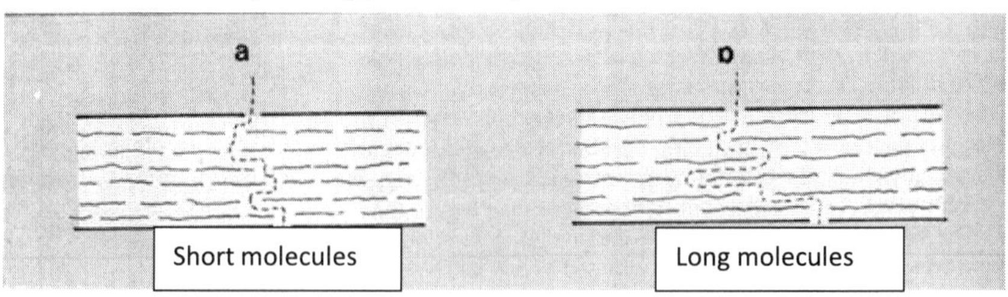

Figure 5.1 Diagram showing the area of attraction in short and long molecules.

5.1.2 Molecular structure of fibres

The properties of the fibre are not only dependent on the individual chains, but also on the way in which these chains are packed together in a fibre (see Fig. 5.2). We can imagine the effect if a number of chain molecules are arranged parallel to one another and if they stick together sideways. The resulting chain bundles show greater rigidity that the individual chains whose flexibility is hidden because sideway attractions prevent the free sliding of chains over one another. It is the primary and the secondary forces, known as valency-bonds, which are responsible for holding the molecules in a chain like formation.

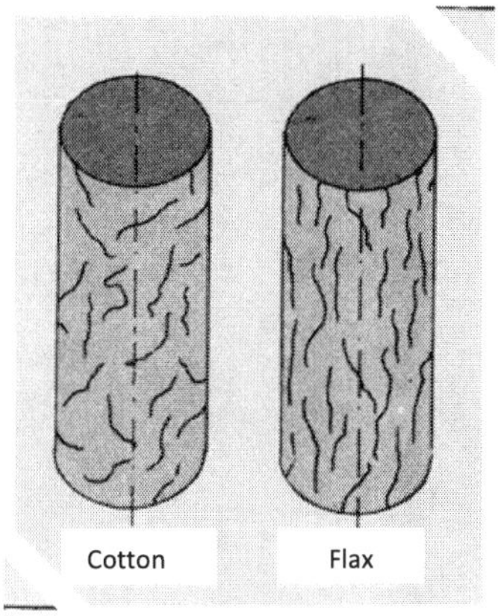

Figure 5.2 Orientation of molecules in cotton and flax.

It is because effects of this kind can occur that we have to take into account the manner of packing together of chain molecules to reach a fuller understanding of the relationship between molecular structure and properties of the fibre. When the chain molecules are randomly arranged, the structure is said to be amorphous. A good example of orientation of molecules in cotton amorphous material is rubber, whose softness and deformability is in direct consequence of a random arrangement of the chain molecules.

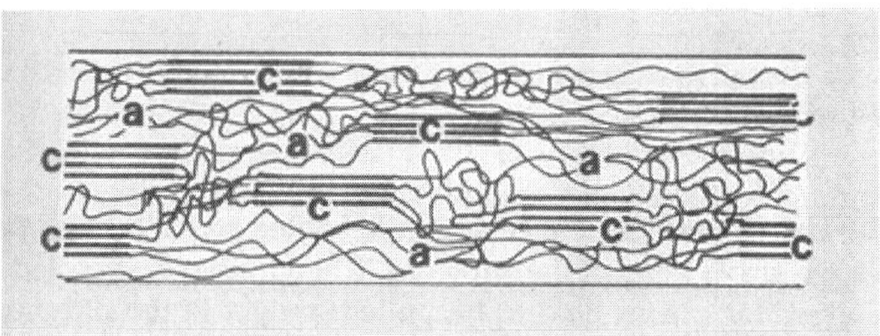

Figure 5.3 Diagram showing crystalline and amorphous regions in molecular structure of fibre a = Amorphous, c = Crystalline.

A very regular arrangement of chains which is in contrast to amorphous structure is the crystalline structure. As a consequence of the regularity of the arrangement, the centres of the secondary forces attract each other so that interchain cohesion is intense. A diagram illustrating the crystalline and amorphous region of the fibre structure is as shown in Fig. 5.3. Between the two extremes of crystalline and amorphous structure, there can exist a series of arrangement with varying degrees of orderliness, and the properties of the fibre will vary accordingly.

Molecular orientation

The molecular orientation of fibres is the alignment of long-chain molecules relative to the fibre axis. The arrangement of the molecules in a natural fibre and the angle of inclination to the fibre axis is a characteristic of each natural fibre and for all practical purposes fixed and unalterable.

Ramie and flax both have extremely high degree of orientation, with high tenacity and low elongation-at-break, whereas cotton has a much lower degree of orientation, with a lower tenacity and higher elongation-at-break.

In the case of manmade fibres it is often possible to control the degree of orientation. The means by which the orientation is increased is invariably that of stretching.

Let us consider a filament which has been extruded without any stretch being applied to it. The molecules will approximate to a random arrangement (as shown in Fig. 5.4) although there will probably be a slight degree of orientation of the molecules parallel to the fibre axis, which has been caused by the direction of flow of the spinning solution through the spinning-jet.

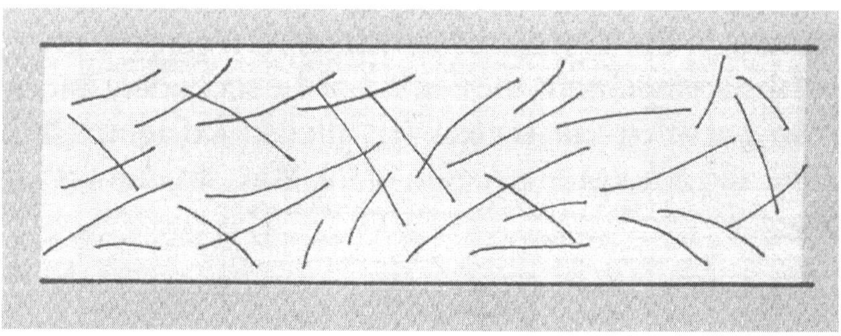

Figure 5.4 Diagram showing random arrangement of molecules before cold (neck) drawing of nylon.

The orderly arrangements of molecules called crystallites are strong and form stable parts of the structure. When a fibre is stretched, the non-crystalline chains tend to be pulled straight in the direction of stretching, that tend to become oriented, parallel to the fibre axis as shown in Fig. 5.5(b). At the same time the crystallites, if not already lying in the same direction are also pulled straight. Hence, the crystallites play a passive role in contributing to the extension of the fibre, since they themselves cannot stretch. When the chain molecules in a fibre are originally fairly well oriented parallel to the fibre axis, there is little scope for extension. On the other hand, when there is plenty of non-crystalline material present, that is, amorphous material present, quite a large extension may be realized before the chains are pulled straight.

The utility of the fibre depends to a large extent upon the amorphous regions within them. Properties depending upon the amorphous regions give the fibre the following advantages:

- To take crease when ironed
- To transmit water vapour when in the form of clothing
- To be tough and flexible
- To recover from large strains
- To permit dyeing and other chemical processing, etc.

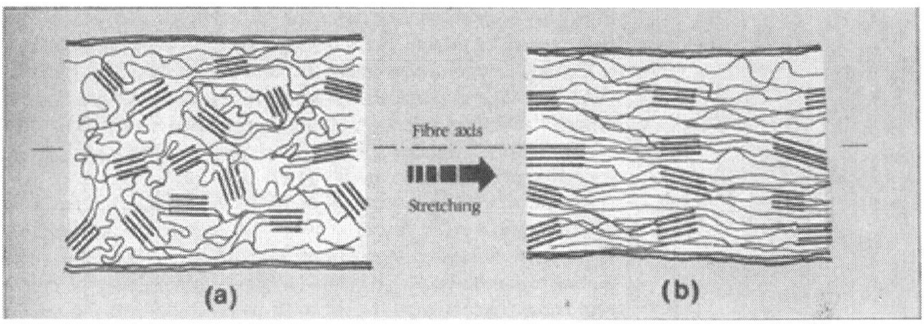

Figure 5.5 Molecular arrangement before and after stretching. Polymer oriented: a = Crystalline polymer not oriented, b = Crystalline polymer oriented.

The mechanical properties of fibres such as, rigidity, crease resistance and extensibility, depend to a large extent upon the amorphous regions in the fibre.

The extension-at-break of textile fibre varies from 2% in glass fibre to as much as 100% for elastic nylon, and the mechanism of extension is quite different in those extreme cases. In general, the extension of a fibre is controlled by the length, orientation and configuration of the chain molecules and the energy content and a number of chemical bonds which hold them together in the amorphous regions of the fibre.

Many artificial fibres possess less strength both in the wet and in the dry conditions. In such cases, increase in the crystalline region in the fibre helps to obtain higher strength. For this reason many artificial fibres are subjected to stretch during their manufacture. The consequence of higher crystallinity and better orientation of the molecular chains are that they increase the robustness of the fibre.

5.2 Fine structure of natural fibres

5.2.1 Fine structure of cotton

In cotton and flax fibres the molecules are arranged spirally along the fibre axis. In the case of flax, the angle of spiral is about 50° and in the case of cotton about 30° to the fibre axis. This spiral structure as shown

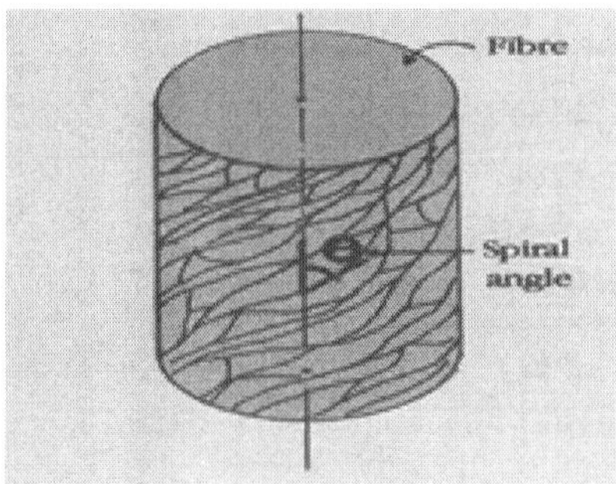

Figure 5.6 Diagram showing the spiral structure of natural fibre.

Fig. 5.6 is common to some of the natural fibres, and is doubtless as characteristic of growth, but it does not occur in the manmade fibres. The spiral structure increases the capacity for extension of fibres. All investigators admit the presence of several concentric layers of different structures in cotton fibre. There are two main structural elements, namely, primary wall and secondary wall. These are arranged concentrically relative to the fibre axis (Fig. 5.7).

The primary wall is a very thin (0.5 mm). It serves as the exterior surface of the fibre. The fibre thickness is mainly formed by the secondary wall having three concentric sections differing in the orientation of macromolecules. The thin layer directly adjoining the primary wall is the outer layer of the secondary wall. Then come to the layers of central part to which the inner layer of the secondary wall is adjoining from the side of the lumen. Inside the fibre there is a lumen with remnants of protoplasm.

Cotton fibre consists of the cellulose fibrils of about 0.025–0.01 mm in diameter in the longitudinal direction. The criss-cross network of fibrils forming primary wall of cotton confer great peripheral strength. Matted arrangement of fibrils of the primary wall provides a tough skin to it.

The secondary wall is almost pure cellulose and accounts for about 90% weight of fibre. Bulk of fibrils arranged alongside each other is aligned as spirals running lengthwise. This exerts maximum strength in longitudinal direction.

Lumen represents considerable volume of unoccupied space containing air cells. Lumen enables cotton fibre to absorb water by capillary action.

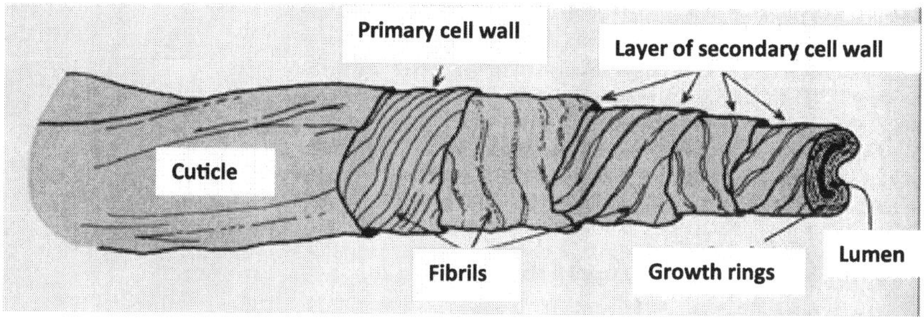

Figure 5.7 Diagram showing the fibre fine structure (microfibrillar) of cotton.

Figure 5.8 Diagram showing the orientation of the cellulose macro molecule.

The orientation of the cellulose macromolecules in the primary wall is low in the central layers of the secondary wall, the orientation of the cellulose macromolecules is higher, here they form long filamentary fibrils inclined at a small angle to the fibre axis (Fig. 5.8).

The composition of the cotton fibre is as given in Table 5.1. For chemical composition and structure, see Appendix I.

Table 5.1 Composition of cotton fibre

Components	Percentage
Cellulose	94.0
Wax	1.2
Sugar	0.3
Proteins	1.3
Ash	1.2
Others	2.6

5.2.2 Fine structure of wool

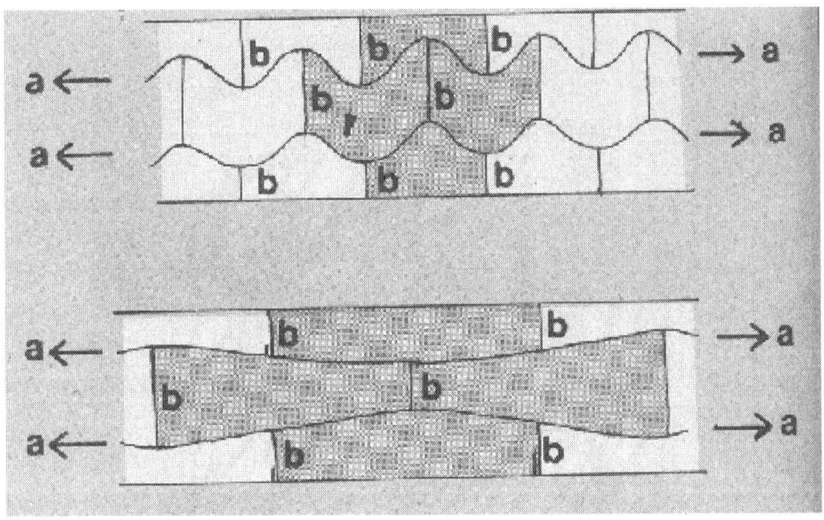

Figure 5.9 Diagram showing the transformation of polypeptide chains from α to β keratin. a: main chain, b: side chain.

Wool fibres are exceptions to the general rule that the crystallites themselves do not stretch when the fibre is extended. In most of the fibres the chain molecules forming the crystallites are already straightened to their full extent, but in keratin the normal state of chains is one of regular folding from which they can be pulled straight. As a consequence of regular folding, the chains can fit together side by side to form a regular pattern. So we have the transformation of keratin from the so-called α-configuration (original fibre) to the β-modification (stretched fibre), and back. In this case, it is characteristic of α-keratin that its main polypeptide chains form regular crimps (folds). On the contrary, the polypeptide chains of β-keratin are almost completely straightened as shown in Fig. 5.9. The main chains of both the first and the second modifications of keratin are interconnected by side linkages which are more or less parallel to a certain molecular plane.

Electron microscopical studies have revealed that the scaly layer consists of three separate fractions, namely, (a) Endocuticle, (b) Exocuticle, (c) Epicuticle. Epicuticle is extremely fine external mantle and is characterized by high chemical stability. The diagram given in Fig. 5.10 shows the relative positions of the components of the cuticle in the cross-section of a non-medulated wool fibre.

The main part of wool fibre is the cortex directly underlying the scaly layer. It consists of separate elongated cells made up of long, slightly flattened and more or less twisted, spindle-shaped cells. The average cells range from 80 to 110 mm in length, 2 to 5 μm in width, and 1.2 to 2.6 μm in thickness.

The spindle shaped cells are disposed in several rows parallel to each other and having their elongated axis directed along the fibre. These cells are constituted of microfibrils which in turn consist of protofibrils.

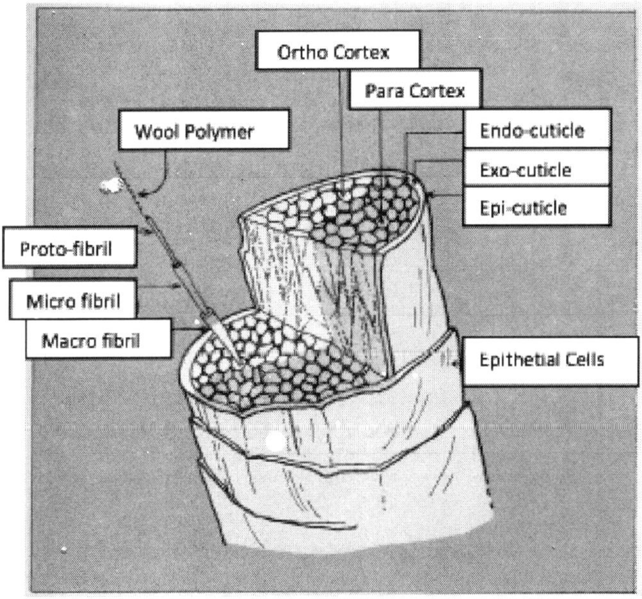

Figure 5.10 Radial section of non-medulated wool showing components of cuticle and cortex.

According to the assumptions which most closely comply with the results of extensive investigations of the wool keratin structure, protofibrils are cables formed of three or two spirals as shown in Fig. 5.11(a).

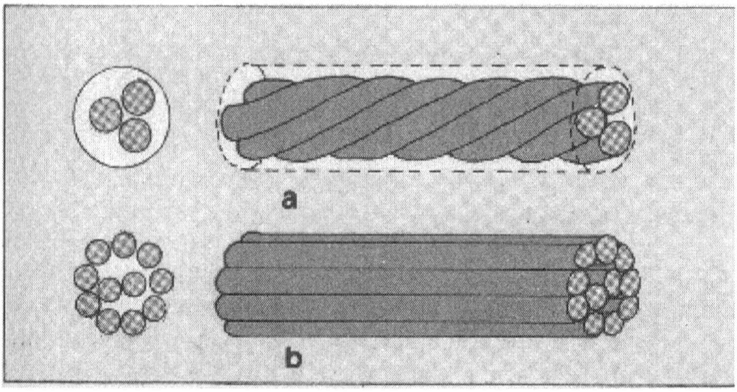

Figure 5.11 Wool keratin structure. a: Protofibrils, b: Mircrofibrils.

Microfibrils consist of eleven protofibrils as shown in Fig. 5.11(b) disposed in a certain order, so that in cross-section, nine of them form an external ring, while two of them are inside the ring. Between the individual spindle-shaped cells there are interlayers of the so-called intercellular substance. Its exact composition is unknown. However, the main elements contained in wool keratin are shown m Table 5.2.

Table 5.2 Composition of wool keratin

Components	Percentage
Carbon	50.3–52.5
Hydrogen	6.4–7.3
Nitrogen	16.2–17.7
Oxygen	20.7–25.0
Sulphur	0.7–5.0

The elemental composition of keratin is nearly the same for different kinds of wool, but the sulphur content varies within great limits. The amount of sulphur is lower in the coarse wool fibres, than in the soft fibres of the same wool.

With keratin the molecular repeat period occurs at intervals of 5.1 Å but when the hair is stretched 100%, the repeat period becomes 10.2 Å.

The presence of many cystine bridges and salt linkages between the chains imparts lateral cohesion and rigidity to the looped formation of the main chain.

The polymer system of wool is extremely amorphous, as it is about 25–30% crystalline and 70–75% amorphous region. Spiralling of the protofibrils, microfibrils and macrofibrils does not imply a well aligned polymer system.

Astbury found that after a short exposure to the action of steam, wool, in a stretched state contracts to less than its original length. Hence the normal unstretched wool fibre is capable of being stretched to twice its original length or contacted to less than its original length. The β-keratin may be regarded as an intermediate form lying between the fully extended and fully contracted modifications, and owing its existence to the crosslinks which inhibit the maximum contraction of the grid.

5.2.3 Fine structure of silk

Silk, unlike the keratin fibre, contains no cystine, but consists chiefly of amino acids with low molecular weights. The polypeptide chains of silk may contain relatively few bulky side chains. This suggests that the segments of polypeptide chains tend to pack closely and considerable interchain hydrogen bonding is possible. Both X-ray and birefringence studies indicate a relatively high degree of orientation and crystallinity of silk fibres. The average composition of raw silk is given in Table 5.3.

Table 5.3 Composition of silk fibre

Components	Percentage
Fibroin	70–75
Sericin	25–30
Mineral substances	1–1.7
Wax	1.5–2.5
Ash	0.8–1

Note: Yellow colouring of cocoons is due to the presence of carotene and Itlll' in a pigment.

The elementary composition of the fibroin of mulberry silk is given in Table 5.4.

Table 5.4 Composition of fibrion

Components	Percentage
Carbon	48.0–49.10
Hydrogen	6.40–6.51
Nitrogen	17.35–18.89
Oxygen	26.00–27.90

5.3 Fine structure of manmade fibres

5.3.1 Fine structure of viscose rayon

Viscose is a linear, cellulose polymer. The molecular structure of rayon is similar to that of cotton or flax, except that the viscose molecules do not have the spiral configuration of the cotton polymer. Also the molecular chains in rayon are shorter and do not form as many crystallites. The breakdown of the cellulose occurs when the alkali cellulose and the viscose solution are aged. When the solution is spun into the acid bath, regeneration and coagulation take place very rapidly. Stretching aligns the molecules to give strength to the filaments. In high wet modulus rayon, since the aging is eliminated, the molecular chains are relatively longer. Because the acid bath is less concentrated, there is slower regeneration and coagulation so that more stretch and more orientation of the molecule can be made. The significant physical differences between rayon and cotton polymer is listed in the Table 5.5.

Table 5.5 Physical differences between rayon and cotton polymers

Polymer	Approximate no. of cellulosic units	Approximate molecular length (nm)	Approximate molecular width (nm)	Approximate degree of polymerization (DP)
Viscose rayon	175	180	0.8	175
Polynosic	300	310	0.8	300
Cuprammonium	250	260	0.8	250
Cotton	5000	5000	0.8	5000

5.3.2 Fine structure of acetate fibres

The secondary cellulose acetate polymer is estimated to be about 40% crystalline and, 60% amorphous, while triacetate or primary cellulose is considered to be somewhat more crystalline. Both these fibres have less crystallinity than cotton.

The important chemical groupings of the acetate and triacetate polymers are the hydroxyl and acetate groups. The former can form hydrogen bonds. These, however, are relatively few in number in the secondary cellulose acetate polymer system, where they do occur. Both fibres relay very heavily upon Van der Waal forces for their intermolecular cohesion.

The acetate or secondary cellulose acetate molecule is about 160 nm long and about 2.3 nm wide, while the triacetate or primary cellulose acetate polymer is about 240 nm long and 2.6 nm wide. Their degree of polymerization range from about 130 for diacetate polymers to about 225 for the triacetate polymers, based upon the cellobiose unit.

Since both types of acetate fibres have molecular backbone of hexagonal units, their polymer could be visualized as disarranged roll of chicken wire. The disarranged portions are in the majority and representing the amorphous regions while the more orderly sections are the crystalline regions of the polymer.

5.3.3 Fine structure of acrylic fibres

The fine structure of acrylic polymer is controversial. The main component of the acrylic polymer is the acrylonitrile monomer. Polyacrylonitrile or the acrylic polymer is a linear polymer, with a degree of polymerization of about 2000. It is made up of molecules about 500 nm long, with width of about 0.3–0.53 nm. The length of the molecules is one of the longest among manmade fibres which are extruded to form textile filaments. This is essential for continuous wet-spinning of the filament from its solution.

The nitrile groups (—C=N) can form strong polar, bonds and the molecules are quite tightly packed. However, due to steric hindrance, the formation of a lattice is not easy and the crystallinity is relatively low. It is believed that a para-crystalline structure is the dominant feature in PAN-fibres.

5.3.4 Fine structure of polyamide fibres

Nylon is a linear, polyamide polymer. The polymer system is estimated to be about 35–55% crystalline. Short inter-molecular distance (0.3 nm) is a result of extensive hydrogen bonding.

The most important chemical group in the nylon polymer is the polar amide group that is –CO–NH– which participates in the hydrogen bonding. The end groups are the amino and carboxyl.

The molecular chains of nylon vary in length. They are 90–140 nm long. Estimated degree of polymerization for nylon 66 is 50–80 and for nylon 6 it is 200. Cold drawing aligns the chains so that they are oriented with the draw direction. High tenacity filaments have a longer chain length than regular nylon.

Unoriented filaments show what can essentially be described as a modified spherulitic structure. This structure has been gradually recognized as a characteristic product of crystallization in polymers is shown in Fig. 5.12. Formation of this kind of structure is common to manufacture of synthetic fibres. During stretching this spherulitic structure is converted into fibrillare structure.

5.3.5 Fine structure of polyester fibre

The polyester polymer is a linear and the most successful fibre in this category is the polyethylene terephthalate fibre. The degree of polymerization ranges from about 115 to 140, resulting in a polymer length of about 120–150 nm, with a width of about 0.6 nm.

The important chemical groups in the polyester polymer are the methylene groups, –CH2–, the slightly polar carbonyl groups, –CO– and the ester groups, –COO–. As the polarity of the polyester is only slight, it is considered to be held together mainly by Van der Waal forces and to a lesser extent some very weak hydrogen bonds. The polyester polymer is estimated to be 30–60% crystalline.

6 Vital fibre properties

The commercial value of any fibre depends largely upon the extent to which it possesses certain qualities or properties, such as length, strength, elasticity, fineness, etc. In this chapter few vital physical and chemical properties are discussed.

6.1 Physical properties

6.1.1 Length of fibres

When the fibres are spun into yarn, they are made to lie parallel and twisted to form a continuous yarn. If the fibres are long there will be fewer ends protruding from the surface of the twisted yarn. Fewer surface end result in a smoother fabric, which has less attraction for dirt particles. The smooth surface also helps in giving an attractive lustre to the finished cloth. Short fibres cause more surface end and more points of potential separation, thus making a relatively weak fabric. The length of the fibre, termed staple length is therefore of fundamental importance, giving rise to the expression: longer the fibre, stronger the yarn. This statement, however, is true only when fibres within a single group are compared.

Long staple yarns within each group of fibres produce stronger, smoother and more serviceable fabrics that cost more than the less durable fabrics made from short staple yarns.

In the case of wool, for example, short, coarse wool is processed on woolen system which is a short sequence process. Coarse variety of yarn is produced to be used in coarse varieties of textiles, e.g. woolen blanket, woolen carpet, woolen fabric, etc. Long fine wool is processed on worsted system which is a long sequence process; fine variety of yarn is produced to be used in fine varieties of textiles, e.g. worsted suitings. A similar distinction is found in every textile fibre.

Cotton fibres are available in the following length ranges:

1. Short staple: 9.52–25.40 mm
2. Medium staple: 12.70–33.38 mm
3. Long staple: 13.75–63.50 mm

The bulk of the domestic cotton crop ranges from 19.05 to 38.10 mm in length.

Wool fibre length varies from an average length of 25.–300.0 mm. Normally worsted yarns are made from the staple fibres of 76.0–200.0 mm length.

Silk fibre from the cultivated cocoons is the longest of the natural fibres. It consists of a continuous fibre which varies in length from 360 to 1200 m.

Jute fibres are from 101.5 to 254 mm long and are cut according to requirements.

All manmade fibres like viscose rayon, cuprammonium rayon and acetate rayon, polyester, polyamide, etc. are produced in continuous filament form. The length is dependent on the ultimate use of the fibres, and on convenience of handling, e.g. fibres are cut to 30, 40, 50, 60, 70 and 80 mm length for spinning on cotton spinning machinery; 70, 80, 90, 100, 120, 150, 180 and 240 mm lengths on woolen and worsted spinning machinery. Since cutting is done with great precision, the staple length is very uniform.

6.1.2 Stress–strain properties of textile fibres

Generally, strength of textile fibres is measured in terms of tensile strength. Tensile properties are characterized by initial Young's modulus, yield point, work of rupture, work factor, elastic recovery, creep, etc. It is therefore worthwhile to understand some of the definitions of the above terminology before proceeding to the subject matter.

6.1.3 Definitions of some basic terms: load

It is the weight (in grams) applied to a specimen which causes a tension to be developed in the specimen.

Breaking load - It is the load at which the specimen breaks, usually expressed in grams weight.

Stress - It is the ratio of the force applied to and the cross-sectional area of the specimen, expressed in dynes/cm^2.

Mass stress - The cross sections of many fibres are irregular in shape and difficult to measure. To simplify matters a dimension related to cross-section is used, such as the linear density of the specimen.

The linear density of a specimen may be expressed in denier or tex unit; then the mass stress becomes the ratio of the force applied to the mass per unit length.

The units of mass stress, therefore, become gram per denier (gpd) or gram per tex (gpt). Here again, abbreviations are used; mass-stress is usually shortened to stress, and quoted in grams per denier (gpd) or grams per tex (gpt) values.

Tenacity - The tenacity of a material is the mass stress at break. The units being of course, grams per denier or grams per tex. An alternative term for tenacity is specific strength.

It is useful to note that by expressing the breaking strengths of different materials in terms of tenacity, comparison can be made directly between specimens of varying fineness. Recommended SI unit for tenacity is millinewtons per tex (mN tex).

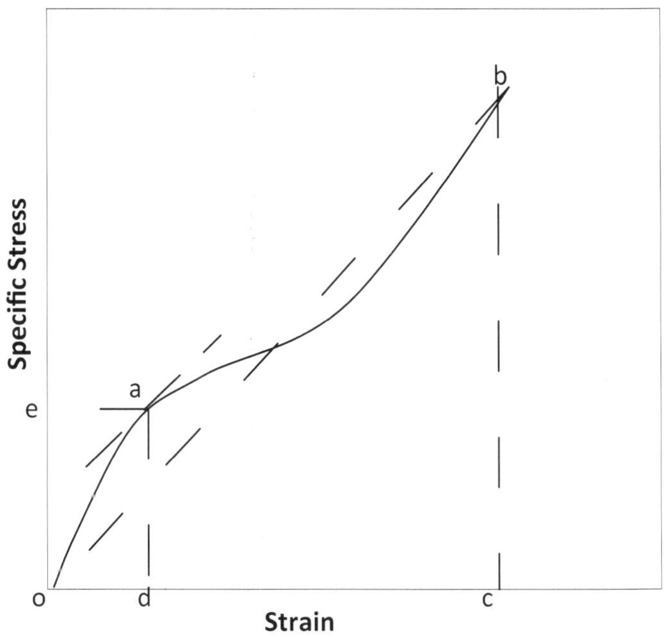

a = Yield point (by Meridith's construction)
oc = Elongation at break
bc= Tenacity
ad=Yield Stress
ae=yield strain

Figure 6.1 Yield point determination.

The count strength product is the tensile strength divided by the weight per unit length of yarn. This quantity has the dimensions of length and is termed the breaking length of a yarn. This is the length of yarn having a weight equal to breaking load that is observed to cause the yarn to break under free-suspension. In other words, it is the length of the specimen which will just break under its own weight when it hangs vertically. Naturally, we do not build tall towers in order to measure this breaking length, but we can calculate it from the results of test on short lengths. The expression of strength in terms of breaking length is useful for comparing the strength of different fibre structures, e.g. for comparing single fibre structure with the yarn strength.

Strain - When a load is applied to a specimen a certain amount of stretching takes place. The amount will vary with the initial length of the specimen. The strain is the term used to relate the stretch or elongation with the initial length. This is expressed in percentage.

Extension - It is the ratio of extended length to the original length of the specimen, expressed as percentage. The extension is sometimes referred to as the strain percentage.

Breaking extension - The breaking extension is the extension of the specimen at the breaking point.

Initial modulus - It is the slope of tangent or the angle between the initial part of the curve and the horizontal axis. It is the ratio of breaking stress to breaking strain. This is expressed as grams per denier or kilometers weight.

Young's modulus gives a measure of the force required to produce a small extension. A high modulus indicates inextensibility and a low modulus indicates great extensibility.

In textiles the term initial modulus is used to describe the initial resistance to extension of a textile material. In other words, this is a measure of fibre's resistance to small initial extension. When the fibre has high resistance to stretching, it will have high modulus. High modulus indicates inextensibility and brittleness. Low modulus fibre requires little force to stretch it. It is also an indication of flexibility.

The initial modulus is a very important part of stress–strain curve. If the initial portion of the curve is straight, it indicates linear relationship between stress and strain. The material behaves like spring and obeys Hook's law. The significance is that when the load is removed the material recovers its original length.

If the stress is in units of grams per denier, then the initial modulus will also be in grams per denier because the strain is a number and is dimensionless.

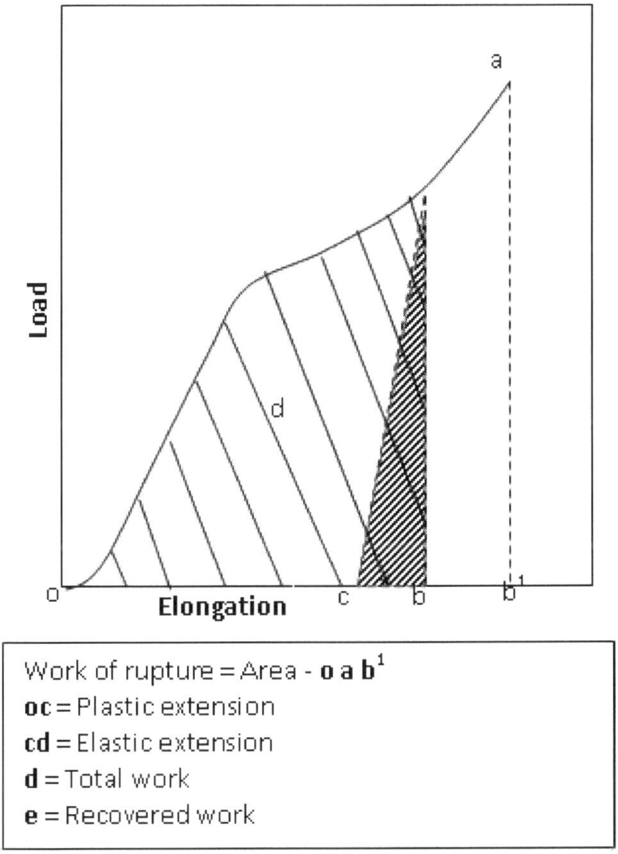

Figure 6.2 Diagram showing work done to break a specimen under tensile load.

Yield point - It is defined as the yielding region, determined geometrically by the point at which the tangent to the curve is parallel to the line joining the origin and the breaking point is taken as the yield point in the Meredith's construction, which is shown in Fig. 6.1

Alternative terms for the yield points are limit of proportionality, 'elastic limits', i.e. where the extension ceases to be proportional to the stress.

Work of rupture - It is the energy or work required to break the specimen, expressed in grams centimetres. Work of rupture is a measure of toughness of the material. It will be proportional to the cross-section of the specimen and to its length. Work of rupture value will indicate the resistance of material to the sudden shock.

If we consider a fibre under load F, increasing in length by an amount dL, then we have:

- Work done by specimen = Force × Displacement
  = F × dL

- Total work done in breaking the fibre = Work of rupture
  = Area under the load = elongation of the curve up to break (Fig. 6.2)

The practical significance of work of rupture will be where the material has to withstand sudden shocks of given energy. Hence, work of rupture is correct quantity to consider in such events as opening of a parachute, a falling climber being stopped by rope and on all occasions when sudden shocks are liable to cause breakage. In tyre cord, belting resistance to elongation is important as it is subjected to constant and repeated application of force.

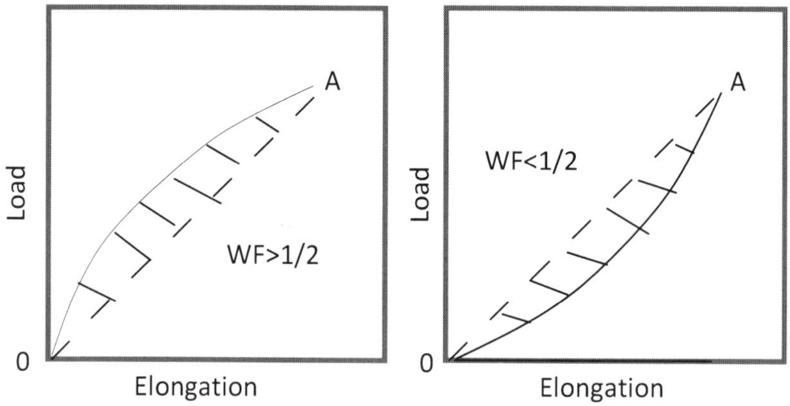

Figure 6.3 Degree of work factor.

Work factor (WF) - Work factor is the ratio of area under the curve, divided by, the breaking stress and breaking strain.

WF = (Area under the curve OA)

(Breaking Stress)(Breaking Strain)

WF = 1/2 when the curve is straight

WF > 1/2 when the curve is convex

WF < 1/2 when the curve is concave

If Hooke's law was obeyed throughout the test, from zero load to zero breaking load, the stress–strain curve would be a straight line therefore be equal to half, i.e. WF = 1/2. Since the textile fibres are not like engineering materials, they do not obey Hooke's law; the stress strain curve obtained will be either concave or convex as shown in Fig. 6.3.

For a curve, which is concave towards the strain axis horizontal the work factor is greater than half, i.e. WF > 1/2.

Conversely, if the curve is concave towards the stress axis, the work factor is less than half, i.e. WF < 1/2.

Elastic recovery - It is defined as that property of a body by which it tends to recover its original size and shape after deformation. The fractional recovery from a given extension may be expressed by the elastic recovery value.

Elastic recovery = (Elastic extension)/(Total extension)

Perfectly elastic materials will have an elastic recovery of 1.0 (i.e. 100%) while perfectly plastic materials will have a recovery of zero (i.e. 0%). Elastic recovery is often expressed as a percentage.

The elastic recovery of different fibres shows considerable variations. Also the same fibre may show variable recovery depending upon the degree of extension and the relative humidity. This can be observed from the data given in Table 6.1.

Load-elongation curve - The load-elongation curve describes the mechanical response of the specimen from zero load and zero elongation up to the breaking point.

From a close study of this curve important information on its mechanical behaviour, like elastic modulus, yield point, tenacity and elongation can be obtained.

Table 6.1 Elastic recovery values of some textile fibres

Elastic recovery (%)						
	1 % Extension		**5% Extension**		**10% Extension**	
Material	**60% RH**	**90% RH**	**60% RH**	**90% RH**	**60% RH**	**90% RH**
Cotton	91	83	52	59	–	–
Wool	99	94	69	82	51	56
Silk	84	78	52	58	34	45
Viscose	67	60	32	28	23	27
Acetate	96	75	46	37	24	22
Nylon	90	92	89	90	89	–
PET	98	92	65	60	51	47
Acrylic	92	90	50	48	43	39

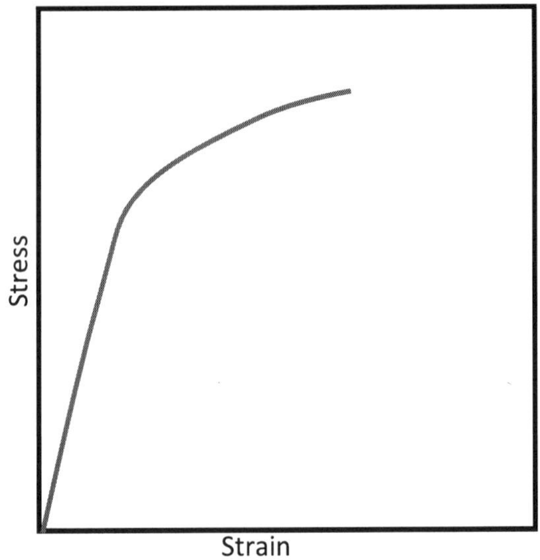

Figure 6.4 Stress–strain curve of a fibre which gives desirable winding characteristics.

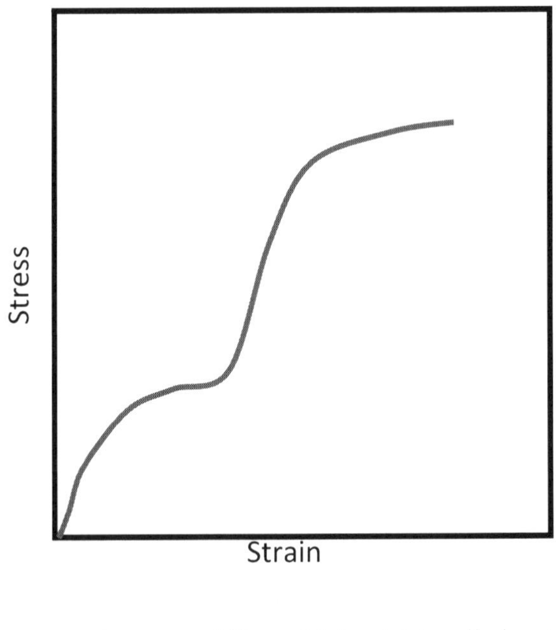

Figure 6.5 Stress–strain curve of fibre which might easily be strained during winding.

Stress–strain diagrams - For some purposes it is more convenient to convert load-elongation curve to a stress–strain curve, thereby enabling more direct comparisons to be made between different types of materials and structures. To convert the curves to stress–strain curves we must draw up a table of values of stress–strain derived from the measurements of load elongation curves. Stress values are obtained by dividing load values by the denier and strain values are calculated by dividing the actual elongation by the test length.

The stress–strain diagram or characteristic curve as it is also called is obtained by using a strength testing machine of autographic type.

The stress–strain behaviour of textile fibre is of great importance in practice, influences to a large degree, the behaviour of fibre in textile manufacture, the interpretation of stress–strain curve assumes great importance.

If the stress–strain curve is a straight line, then the fibre is said to obey Hooke's law and so long as the curve is straight, the fibre will be elastic. Usually, after a small extension, the elongation of fibre becomes greater than that required by the straight line curve. This generally means that plastic flow is setting in, and that the fibre is no longer truly elastic. The yield point corresponds to the point where the curve departs from the straight line behaviour. For most of the fibres this occurs at fairly low strains.

A fibre which will be most satisfactory for textile purposes is one that does not stretch much for small loads, so that during winding it does not stretch appreciably. The advantage of this is that even if the winding tension varies slightly from spindle to spindle, the yarn will not be susceptible to high stretches for very small loads. Consequently a stress–strain diagram, which for small loads up to, say 0.2 or 0.3 gpd, is straight, is desirable. Two characteristic stress–strain I diagrams are given in Figs. 6.4 and 6.5 to illustrate the significance, of the stress–strain curve.

A fibre which gave the stress–strain diagram shown in the Fig. 6.4 would, other things being equal, be preferred to another fibre which gave the curve shown in Fig. 6.5.

It is usually found that the fibres showing curve of Fig. 6.4 type have a better and softer handle, and they drape better.

From the stress–strain curve given in Fig. 6.6, it is clear that the cellulose acetate has more pronounced influence than viscose rayon and is softer to feel. Wool and the regenerated fibres (Casein, Ardil) have the best handle, and they all have a low initial modulus and a high compliance ratio, which appears on the characteristic curve as a flattening for moderate extensions. It is probably true to say that the fibres with the softest handle are those that are most easily damaged by excessive winding tensions.

The fibre that is the best for soft and comfortable underwear is unlikely to be the best for industrial uses such as ropes, belts, where high strength and good recovery are essential.

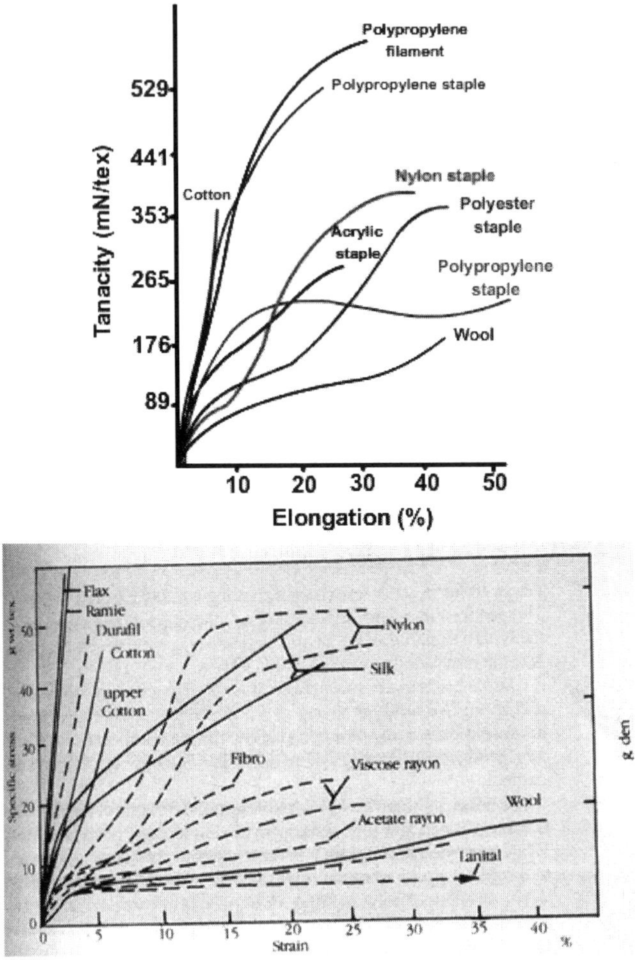

Figure 6.6 Stress–strain curves of some important fibres.

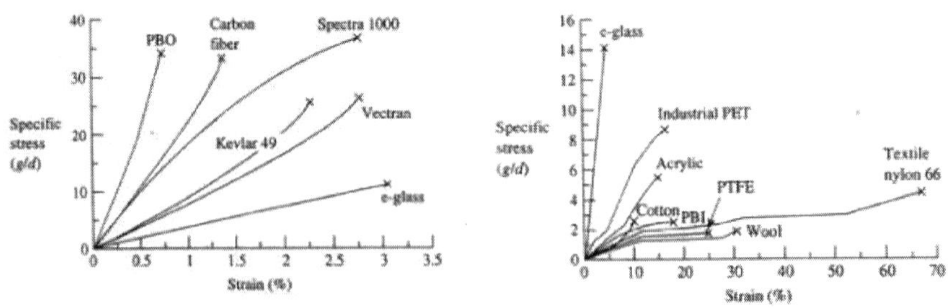

Figure 6.7 Stress–strain curves for high performance fibres.

Creep - The total extension in a specimen is composed of two elements: the instantaneous extension and the time dependent effect. The latter is called creep.

On the removal of the load the specimen recovers rapidly at first and then more slowly with perhaps a small amount of residual extension which is called permanent set. Similarly, the time dependent effect may be considered in two parts, the primary creep which is recoverable and the secondary creep which is not recoverable.

6.1.4 Tensile strength of fibres

Fibres differ in their resistance to tearing when subjected to tension. The serviceability and durability of a fabric cannot be determined by the length of staple alone. Tensile strength is a prime necessity in fibres if resultant fabric is to stand strain of wear.

Glass fibre is a very strong fibre. It is about twice as strong as cotton and about five times as strong as wool. Still its use is limited, main because it lacks many other properties which are important for consumer products. Tenacity of fibre glass is 6.3–6.9 gpd (grams per denier).

Polyamide is noted for its high strength. The tenacity of regular nylon is 4.6–5.8 gpd and high tenacity nylon has tenacity of about 8.8 gpd. While wet nylon retains much of its strength, the wet tenacity may drop to 4.0 5.1 gpd for regular nylon and to 7.6 for high tenacity fibre.

The strength of polyester fibre varies from a low value of 2.5 gpd for the same staple fibre, to a high value of 9.5 gpd for some high tenacity fibre. There is no loss of strength when polyester fibres are wet.

Silk is the strongest of all the natural fibres, being nearly equal to that of iron wire of the same diameter in terms of specific strength. It has a tenacity of 2.4 5.1 gpd when dry, and its wet strength is about 80–85% of the dry strength.

Flax is a strong fibre with a tenacity of 5.5–6.5 gpd. Occasionally, some fibre of inferior quality may have tenacity as low as 2.6 gpd. When wet the fibre is even stronger than dry state with an increase in strength of about 20%.

Jute has a tenacity of about 3–5 gpd. Cotton has a tenacity of 3–5 gpd. When wet, the strength of the cotton increases by about 25%.

The tenacity of acrylic fibre varies between 2 and 4.2 gpd. Wet tenacity of the acrylic fibre ranges from 1.6 to 3.8 gpd.

The dry strength of viscose rayon of regular tenacity is 1.5–2.4 gpd and that of the medium and high tenacity fibres ranges from 2.4 to 3.2 gpd 3.0 to 5.0 gpd, respectively. The new high tenacity and high wet modulus viscose fibres are considerably stronger than earlier types have wet and dry strengths (2.7–4 gpd and 3.4–5.5 gpd, respectively) equal to or better than cotton.

Cuprammonium rayon fibre, however, is not produced in medium or high tenacity forms. This fibre possesses dry tenacity of 1.7–2.3 gpd, and wet tenacity of 0.95–1.35 gpd. Acetate rayon is relatively weak fibre. Acetate has a dry tenacity of 1.3–1.5 gpd. When wet, the strength drops to approximately 0.9 gpd. Amel, the triacetate fibre possesses a dry tenacity of 1.2–1.4 gpd and a wet strength of 0.8–1.0 gpd.

Wool has the lowest tensile strength of all the natural fibres. The strength of wool is 1.0–1.7 gpd when dry. When wet, the strength drops to 0.8–1.6 gpd.

Table 6.2 Tensile properties of some textile fibres

Fibre	Tenacity (N/tex)	Breaking extension (%)	Work of rupture (mN/tex)	Initial modulus (N/tex)	Work factor
Cotton					
StVincent	0.45	6.80	14.90	7.00	0.49
Upper	0.32	7.10	10.70	5.00	0.46
Bengals	0.19	5.60	5.10	3.90	0.49
Flax	0.54	3.00	8.00	18.00	0.50
Jute	0.31	1.80	2.70	17.20	0.50
Hemp	0.47	2.20	5.30	21.70	0.50
Ramie	0.69	3.70	10.60	14.60	0.47
Viscose rayon – continuous filament (Courtaulds)	0.18	27.20	30.60	4.80	0.62
Fibro	0.21	15.70	18.80	6.50	0.59
Tenasco	0.27	16.90	19.70	6.00	0.60
Acetate (Celanese)	0.13	23.70	21.60	3.60	0.72
Fortisan (Cellulose)	0.59	6.40	19.10	16.10	0.51
Silk	0.38	23.40	59.70	7.30	0.66
Nylon	0.47	26.00	76.00	2.60	0.61
Wool					
Botany 64s	0.11	42.50	30.90	2.30	0.64
Crossbred 56s	0.14	42.90	37.50	2.10	0.62
Crossbred 36s	0.12	29.80	26.60	3.00	0.74

Glass	0.75	2.50	9.80	29.40	
Steel wire	0.26	0.8	17.7	28.5	

6.1.5 Elasticity of fibres

Elasticity is the property of a body by virtue of which it tends to recover its original shape and size after deformation.

Elasticity includes several factors, like elastic recovery, elongation and resiliency. Elastic recovery is the per cent of return from elongation towards the original length. Elongation is the amount of stretch or extension that a fibre will accept. Breaking elongation is the amount of stretch that occurs to the point where the fibre breaks. Elastic recovery is an important factor in the resiliency of a fibre and usually good elastic recovery indicates good resiliency.

When a fibre is subjected to a force, it will stretch to certain degree. When it is stretched by a small amount it may exhibit almost perfect recovery. But when it is stretched to a greater degree it may react in much more complex way. The extent, to which fibre becomes permanently deformed when it is stretched, is of great technical significance.

The amount of elongation in a fibre is an important factor in evaluating elastic recovery. Some fibres with low elongation have excellent elastic recovery; however, this elastic property is of little value because elongation and elasticity need to be considered together in fibre evaluation. A fibre with extremely high elongation but medium to low elastic recovery level might be undesirable because the product could not return to its original size after extension.

Elasticity is very important in textiles for the general reaction of fibre to longitudinal tensions and to flexing backwards and forwards. It has an immense and decisive influence on the properties of end products made from fibre.

Elastic recovery and dimensional stability are of paramount importance in textile clothing. For example, a suit made from a material with poor elastic properties would very soon exhibit baggy knees in trousers and baggy elbows in the jacket. Crease resistance of a fabric is usually a consequence of the resiliency of the fibre itself.

Crimp, the waviness in a fibre, increases elasticity and elongation. It varies depending upon the degree of the crimp a fibre possesses. Wool is the only natural fibre which possesses noticeable crimp. In fine quality wool there will be about 30 waves per 25 mm; and in coarse wool there will be 5 to 10 waves per 25 mm.

If crimps are absent in natural fibres, artificially permanent crimps cannot be introduced to them. Permanent crimp can only be introduced to the thermoplastic fibres by a process known as texturing, whereby the elasticity of the fibre undergoes change.

According to this broad definition of elasticity, it is difficult to establish an absolute range for the fibres. The following generalities are adequate for average consumer needs.

Nylon is highly elastic fibre with relatively good elongation. The elastic recovery properties assure excellent shape retention of nylon fabrics. Nylon has good-to-very good recovery from creasing or wrinkling, the same can be observed in Table 6.3.

Table 6.3 Breaking elongation and elastic recovery values of nylon 6 and nylon 66

Fibre type	Breaking elongation (%)		Elastic recovery at 4% elongation
	Dry	Wet	
Nylon 6	23–45	28–47	99–100
Nylon 66	26–65	30–70	100
Nylon 6			
High tenacity	16–20	19–33	99–100
Nylon 66	16–28	18–32	100

Wool is one of the most elastic fibres. It can be stretched to about 30% of its original length without breaking. Wool has elastic recovery of about 99% at 2% elongation and 63% at 20% elongation and elongation-at-break is about 25–35% in dry state and 25–50% when it is in wet state.

Silk possesses a very slow elastic recovery. Acrylic fibres have approximately the same elasticity. They do not have stretchability, but they have good recovery of about 99% at 2% elongation. Elongation-at-break (dry strength) ranges from 20 to 30% depending on the type of acrylic, e.g. Orlon staple 20–88, Acrilan staple 34–50. Elongation-at-break in wet state ranges from 26 to 60%, i.e. 34–60 for Acrilan staple and 26–34 for Orlon staple.

All the acrylic fibres are very resilient; they do not get wrinkled easily, and any wrinkles that may form fall out readily.

Polyester fibres which are stronger have lower degree of elongation. Elastic recovery of high tenacity and staple fibres is very good. The regular tenacity polyester exhibits some variation in elastic recovery (see Table 6.4) but they are superior to cellulosic fibres and only slightly inferior to nylon.

The property of resilience and recovery from creasing and wrinkling is excellent in polyester fibres. Furthermore, heat setting can stabilize fibres and yarns, so they need little or no pressing.

Table 6.4 Elongation-break and elastic recovery of polyester fibre

Fibre type	Elongation-at-break (%)		Elastic recovery percentage
	Dry	Wet	
Polyester (terylene)			
(i) Regular tenacity filament	19–30	19–30	97 at 2%
			80 at 8%
(ii) High tenacity filament	10–14	10–14	100 at 1%
(iii) Staple filament	18–50	18–50	100 at 1%

Acetate rayon has fair stretchability as well as recovery to original size. It also has fair resilience. The elongation-at-break when dry is about 25–45% and about 35–50% in wet state. Elastic recovery value is about 48–65% at 4% elongation. Acetate rayon fabrics will wrinkle, but not as easily as viscose rayon or cotton, and there is a tendency for the wrinkles to fall out.

Viscose and cuprammonium rayons have approximately equal properties in this category. This can be seen from Table 6.5.

Elasticity of viscose rayon and cuprammonium rayon is fair. They are not very much wrinkle resistant. To overcome this, viscose, rayon fabrics can be processed to impart a good degree of wrinkle resistance. Spun rayon fabrics tend to be more naturally wrinkle resistant because of their soft texture.

Table 6.5 Elongation-at-break and elastic recovery values of viscose rayon fibres

Fibre type	Elongation-at-break (%)		Elastic recovery percentage
	Dry	Wet	
Viscose rayon			
(i) Regular tenacity	15–30	20–40	30–74 at 4
(ii) High tenacity	9–26	14–34	97 at 2
Cuprammonium Rayon	7–23	16–43	17–45 at 4

Cotton has little or no elasticity compared to polyamide, wool, etc. Therefore, fabrics made from cotton wrinkle easily. When a piece of cotton fabric is creased with finger nail the crease is retained. Silk and wool are not so easily affected. Finger nail creasing may serve as a test, therefore, to distinguish between silk and wool from cotton and linen, cotton fabrics can be processed to be wrinkle resistant.

Jute and flax are the least elastic and resistant of the natural fibres. Lower elasticity gives jute an advantage as bagging material as it means that loaded bags have less tendency to shift about when stacked than bags made of fibres with greater percentage of elongation.

Linen fabrics wrinkle and crease easily; however, they can be processed to improve their wrinkle resistance.

Fibre glass is brittle, and lacks pliability, and although it is often described as elastic, the word is not used in the customary meaning. Fibre glass has elongation of 2–4%.

6.2 Thermal properties

It is important to know what effect heat has on textile fibres, whether a fibre is inflammable, and how a fabric is to be ironed without being melted or disintegrated.

Cotton will scorch if ironed with too hot temperature. This causes damage and loss of strength. But the ironing temperature of cotton is relatively high before change occurs. Safe ironing temperature for cotton is 218 °C.

The reaction of linen fibre to heat is similar to that of cotton. Safe ironing temperature for linen fibre is around 232 °C.

Wool scorches rather easily and becomes shiny with pressure; it should be pressed at low temperature on the wrong side of the fabric with a pressing cloth. Scorches damage the wool permanently, discolouring it, and making it harsh. Safe ironing temperature for wool is 300 °C. Silk is less resistant to heat than wool. Safe ironing temperature is same as that of wool. Silks should be ironed damp under a press cloth. Wild silks should be ironed dry.

Viscose and cuprammonium rayons burn rapidly, once the kindling point of about 177 °C is reached. A slightly cooler iron is recommended for these fibres than for cotton. Acetate rayon sticks to the iron at about 177 °C. At higher temperature it will char, sputter and flame. A cooler iron should be used for acetate than for rayon. Cigarette sparks and glowing embers falling on acetate fabrics will melt holes in the fabric on contact.

Nylon will melt at 250 °C. It may, however, discolour or yellow at lower temperatures. It is advisable to iron nylon in the same manner as acetate. Nylon garments may be commercially heat set to retain their shape after washing or dry cleaning without further pressing or ironing. As with acetate fabrics, cigarette sparks and glowing embers will melt holes in nylon fabrics.

Polyester melts at about 250 °C. It should be ironed in the same manner as nylon. These fibres can be heat-set to hold a permanent shape or crease.

Orlon melts at about 235 °C. It has thermoplastic heat setting properties similar to polyester. Almost all acrylic fibres do not support combustion till 150 °C. Some of them soften and shrink above this temperature.

Fibre glass begins to lose some strength at 315 °C and softens at 815 °C. Obviously, ironing not only will affect fibre-glass but is actually a waste of time and effort.

6.3 Biological properties

Many microorganism attack fibres weather natural or manmade. Numerous fungi cause mildew, with discolouration from pigments in the organisms, weakening or rotting of the fibre, and a musty smell. Certain bacteria also cause micro-biological rotting and appear to be its main cause under waterlogged conditions. Most fungi reproduce by means of spores, largely distributed through the air, and present wherever fibres are found.

Cotton is damaged by fungi, such as mildew and bacteria. Mildew will produce a disagreeable odour, and will result in rotting and degradation of cotton. Certain bacteria encountered in hot, moist and dirty conditions cause decay of cotton. Moths and beetles do not attack and damage cotton. Silverfish will eat cotton cellulose.

Flax has excellent resistance to mildew, but if it is in a moist condition or in a very humid atmosphere, mildew grows rapidly and damages the fibre. Other bacteria will not attack flax unless the fibre is exposed to severe conditions of dampness. It is generally resistant to most household insects and pests.

Jute has good resistance to microorganisms and insects. Moisture increases the speed of deterioration but dry jute will last for a very long time.

Rayons of all types are destroyed by mildew. High wet modulus viscose tends to exhibit their resistance to these bacteria.

Wool has good resistance to bacteria and mildew. However, both organisms may attack stains left on wool. If wool is stored in an atmosphere where moisture is present, mildew will form and over a period of time it will destroy the fibre. Wool is an appetizing meal for several types of insects. It has been estimated that these insects damage several million pounds of wool fabric each year.

Silk is resistant to attack by mildew and is relatively resistant to other bacteria and fungi. Silk has good resistance to the clothes moth, but carpet beetles will eat it.

Synthetic fibres like polyester, polyamide, acrylic, etc., are highly resistant to attack by most insects or microorganism.

6.4 Chemical properties of fibres

6.4.1 Effect of acids and alkalies

Knowledge of reaction of textile fibre to the alkalis and acids is important, as alkalies destroy animal fibres; and acids, even in a weak solution, destroy vegetable fibres. Such chemical tests are valuable in determining the fibre content and percentage blend composition of fabrics. Good results in the bleaching, finishing and dyeing processes are dependent upon the reaction of the fibres to alkali or acid chemical agents. Not only is this reaction important in testing and manufacturing, it also closely touches the consumer when the care of clothing is considered.

The reaction to alkalies and acids is most important to the consumer when fabrics are washed. As the brown soap sometimes used for household laundry is strongly alkaline, it must not be used for silk or wool fabrics. Synthetic detergents, which are neutral, should be used for these protein fibres. For cotton and linen, synthetic detergents are not as effective as soap.

Cotton is destroyed by concentrated inorganic acid such as hydrochloric, sulphuric and nitric, and is damaged when such acids are in dilute solutions. It is not affected by alkalies even in hot, strong solutions. Strong soap may be used for washing cotton.

Reaction of linen to acids is similar to that of cotton. Linen is not readily damaged by alkalies, but strong soap should not be rubbed on the fabric as it will turn linen yellowish.

Wool is destroyed not by dilute acids, but by concentrated acids. Wool is quickly destroyed by strong alkalies. It dissolves completely if immersed in a boiling alkali. A neutral or mild soap must be used for wool.

Silk is affected a little by organic acids. Weak inorganic acids also have a little effect, but strong solutions injure silk. A dilute solution of nitric acid produces a bright yellow colour in silk, and this reaction identifies silk from wool. Silk is destroyed by strong alkalies. Strong soap should be avoided for washing silk. Compared with wool, silk is much resistant to alkalies but little resistant to strong acids.

Rayons are not readily damaged by alkalies, but are less resistant than cotton and linen. Both viscose and cuprammonium rayon lose tensile strength when exposed to a weak solution of a strong alkali, such as caustic soda. Hot dilute solutions or cold concentrated solutions of strong inorganic acids, such as sulphuric, hydrochloric and nitric, destroy rayon.

The inorganic acids, formic and acetic, do not affect viscose or cuprammonium rayon. Strong acids will destroy acetate rayon. Acetate rayon will, however, withstand solutions of organic acids as strong as 28%. Acetate rayon is easily affected by alkalies, especially, if the water is very hot. Mild soaps and lukewarm water should, therefore, be used when laundering acetate.

Nylon is affected by concentrated solutions of the strong mineral and organic acids. Alkalies, however, have virtually no effect on nylon.

Polyester has good resistance to most of the mineral acids, although a concentrated solution of sulphuric acid will decompose it. Polyester has good resistance to weak and strong alkalies; however, it will be disintegrated by strong alkalies at boiling temperatures.

Acrylics have good-to-excellent resistance to mineral acids and fair-to-good resistance to weak alkalies. Acrylics are not harmed by the organic acids.

Fibre glass is affected by hydrofluoric and hot phosphoric acids only. It is damaged by hot solutions of weak alkalies and cold solutions of strong alkalies.

6.4.2 Effect of sunlight on fibres

All fibres are affected by the radiation in sunlight which brings about deterioration following prolonged exposure.

When textile fibres are exposed to light either natural sunlight or light from an artificial source, it normally suffers progressive decrease in strength and elongation. In both types of exposure, criteria of degradation may be expected to show some qualitative agreement. Results differ quantitatively due to great differences in the distribution of spectral energy in two types of exposure.

Photochemical degradation is an acceleration of a normally very slow oxidation of say, e.g. cellulose by oxygen in surrounding atmosphere. Reaction proceeds more rapidly when moisture is present.

When a fabric is surrounded by inert atmosphere, cotton is not significantly degraded, cellulose irradiated by ultraviolet light in an inert atmosphere was not degraded but it was sensitized to the extent that rapid degradation would occur if subsequently exposed to oxygen in the absence of irradiation.

Cotton fibre, when subjected to a prolonged exposure to sunlight will cause the fibre to turn yellow and gradually cause degradation. This damage is accentuated in the presence of moisture.

Improperly stored, cotton retains most of its strength and appearance over a long period of time. Cotton should be stored in dark, dry areas.

Linen shows a gradual loss of strength when it is exposed to sunlight, but this is not serious. Flax-linen makes good choice for curtain and drapery fabrics. If stored properly, linen will age remarkably well. Table coverings and sheets packed away for very many years have proved to be as strong as the new fabrics.

Rayon fibres undergo deterioration when exposed to the ultraviolet rays of sunlight. Rayons have good resistance to ageing and, if properly stored will last for many years.

Wool fibre has better resistance to sunlight than cotton. Silk tends to accelerate the decomposition when it is exposed to sunlight.

Sunlight has a destructive effect upon nylon, and there is a marked loss of strength after extended exposure. Left in the direct sun for several weeks, nylon may decompose.

Polyester exhibits good resistance to sunlight when it is kept behind glass, but if it hits directly by sunlight, the fibre is weakened after prolonged exposure. Polyester fibres have excellent resistance to possible damage due to ageing like polyamides.

Acrylic fibres have excellent resistance to sunlight and all other climate elements. Age does not appear to affect acrylics.

Both polyethylene and polypropylene loose strength and slowly degrade following prolonged exposure to sunlight.

7
Moisture relations

Some of the most important properties of a textile fibre are closely related to their behaviour in various atmospheric conditions. Most of the fibres are hygroscopic or hydrophilic, that means, they are able to absorb water vapour readily from moist atmosphere and, conversely, desorb or loose water in dry atmosphere. Very few fibres belong to the class of heteroscopic or hydrophobic nature. These fibres are not able to absorb water vapour readily from moist atmosphere. Many physical properties of a fibre such as tensile strength, elastic recovery, electrical resistance, rigidity, etc., are affected by the amount of water absorbed. When in fabric form, the moisture absorbency of a fibre plays a major part in deciding whether the fabric is suitable for a particular purpose or not.

Mechanism of water absorption is a subject beyond the scope of this book, but it is worthwhile to know some of the terminologies used in this connection. They are as follows:

7.1 Definitions of some important terms

7.1.1 Moisture regain

Moisture regain is defined as the weight of water in a material expressed as a percentage of its oven-dry weight.

7.1.2 Moisture content

Moisture content is the weight of water in a material expressed as percentage of its total weight.

The amount of moisture in a sample of material may, therefore, be expressed in terms of moisture regain percentage or moisture content percentage. Hence, the relationship between the above terms is considered as follows:

Moisture region (%) = Weight of water (W) in the sample

Oven-dry weight (D) of the sample × Weight of water in the sample × 100% or M = 100 W/D

The relationship between moisture regain and moisture content is:

$R\,\% = M + 1 - (M/100)$

$M\% = R + 1 + (R/100)$

For the commercial purpose, it is essential to know the official standard regain value of each fibre. Table 7.1 shows both, the official standard regain (*British Standard Handbook,* 1956 Ed) and the standard regain.

Table 7.1 Moisture regain values of textile fibres

Fibre type	**Moisture regain (%)**	
	Official region value	**Standard regain value at 65 RH 20 °C**
Cotton	8,5	7
Wool	14 - 18.25	11 to 16
Silk	11.0	8
Jute	13.75	11-13
Viscose rayon	13.00	1143
Cuprammonium rayon	11.0	12.5
Acetate rayon	6.0	6.0
Nylon 6	4,5	2.8 to 5
Nylon 66	4.5	2.8 to 4.5
Polyester	0.4	0.4{0 0.8
Acrylic	1.5	1.5

Hysteresis effect - Suppose two samples of the same material were taken into a given atmosphere, one completely wet and the other dry, and at intervals of time, the regain values for each were determined by plotting the regain against time. For both samples, the two curves would be obtained which are shown in Fig. 7.1. The regain changes fairly quickly at first and then more slowly as equilibrium conditions are approached. One might expect that the two curves would meet, but curiously enough the equilibrium regain values differ. The sample, which was originally wet, has a higher regain value than the other, an effect known as hysteresis.

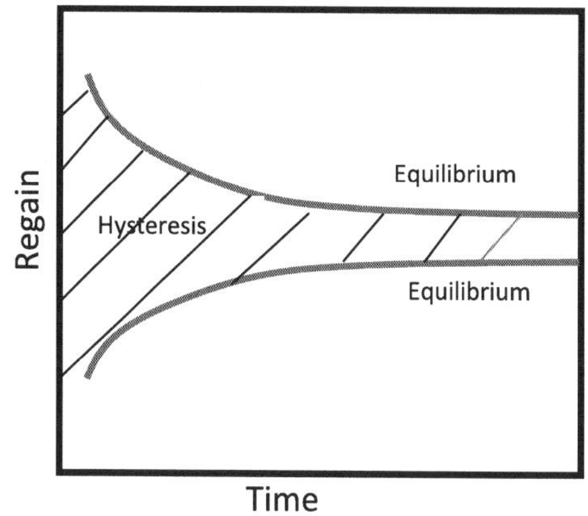

Figure 7.1 Moisture regain vs time

7.1.3 Absolute humidity

Absolute humidity is defined as the weight of water present in a unit volume of moist air.

7.1.4 Relative humidity

Relative humidity is defined as the ratio of the actual vapour pressure to the standard vapour pressure at the same temperature expressed as percentage.

Since the relative humidity affects the moisture regain of a textile material, and since the properties of the material are influenced by the regain (Fig. 7.2), it is necessary to specify the atmospheric conditions in which the testing should be carried out. The standard conditions suggested for testing textile materials are 65% ± 2% RH and 20 ± 2 °C temperature and for the tropical and subtropical regions 65% ± 2% RH and 27 ± 2 °C temperatures.

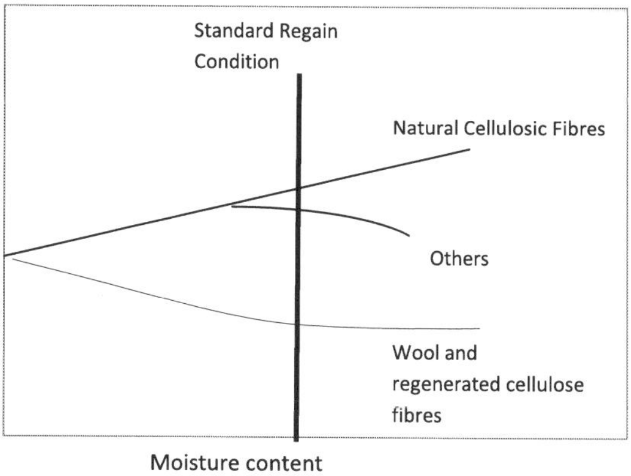

Figure 7.2 Effect of moisture content on tensile strength of fibres.

7.2 Some effects of regain on fibre properties

7.2.1 Dimensional changes

Absorption of moisture is accompanied by changes in the dimensions of fibres. Swelling is mostly transversal, since water molecules penetrate between the more or less parallel molecular chains and exert their forces outwards. The length of the fibre increases due to absorbed moisture. The net result of absorbed moisture in fabrics is shrinkage in length which is, again, a function of the structure of the fibre assembly rather than the inherent fibre behaviour.

Dimensional changes in the fabric, as it passes from one atmosphere to another, may result in the wrinkled appearance of suits tailored in one atmosphere and worn in others, especially in climates where higher humidity is found.

Advantage of swelling is taken in the design of water proofs. A close weave can be used which becomes even closer as the component yarns swell, thus, preventing the penetration of water.

7.2.2 Mechanical properties

The general effect of water molecules in the fibres is to reduce the size of the forces holding the molecular chains together, thereby weakening the link (with the exception of vegetable fibres). Not only is the maximum breaking strength reduced, but the stress-strain curve assumes a different form. The yield point may be shifted, and applied loads, which on dry fibres would not cause damage, may stress the fibres beyond the lowered yield point. Other mechanical properties affected by regain include extensibility, crease recovery, flexibility, etc.

7.2.3 Electrical properties

The "before and after" effect of moisture on the electrical resistance of textile materials is most striking. The ratio of the resistance at low regain and at high regain can be of the order of hundreds or thousands to one, circumstance which has led to the design of moisture meters based on the measurement of resistance value of textile materials.

Other electrical properties affected by the amount of moisture in the material are the dielectric characteristics and the susceptibility to static troubles. Changes in dielectric characteristics are a source of error in the measurement of irregularity of slivers, rovings, and, yarns on capacity type testers.

7.2.4 Thermal effects

When moisture is absorbed by textile materials, heat is generated. This heat is referred to as the "heat of absorption" or "heat of wetting". For example, in passing from an atmosphere of 45% RH and 18 °C temperature to an outside atmosphere at 95% RH and 50 °C, a man's woolen jacket weighing 1 kg will produce 100,000 calories of heat, i.e as much heat as normal body metabolism will produce in one hour.

From the theoretical and practical point of view, heat sorption (W) assumes importance. This is defined as the heat evolved by the fibre when one gram of water is absorbed into a large mass of it.

From the experiments it has been found that wool generates more heat than other fibres while heat generated by nylon is the smallest. Therefore, the practical advantage of wool fibre in keeping a person warm by such a mechanism is apparent. Table 7.2 shows the heat of wetting values for some important fibres.

Table 7.2 Heat of wetting of dry fibres from zero_regain

Fibre type	**Heat produced (cal/g)**
Wool	26.9
Viscose rayon	25.2
Cotton (mercerized)	17.5
Silk	16.5
Flax	13.0
Ramie	11.1
Cotton (normal)	11.0
Acetate	8.2
Nylon	7.3
Acrylic (orlan)	1.7
Polyester (terylene)	1.2

7.3 The effect of relative humidity on strength and other properties of textile fibres

The tensile behaviour of textile fibres depends upon such factors as temperature and time, besides the relative humidity of the surrounding atmosphere. It is necessary to keep the temperature constant while studying the effect of humidity on stress-strain curves of various textile fibres.

The stress-strain curve drawn in Fig. 7.3 shows the effect of relative humidity on strength and related fi bre properties. These curves were obtained at temperature of 200 °C and at such a rate of extension that the rupture occurred within 10 to 30 sec.

From the curve given in Fig. 7.3, it is evident that as the relative humidity increases, all the fibres become more extensible, i.e., the stress-strain modulus, i.e., ratio of stress to strain becomes smaller and the extension at break becomes larger This is due to the absorbed water reducing the cohesion of the chain molecules in the amorphous region of the fibre leaving the crystalline regions unaffected

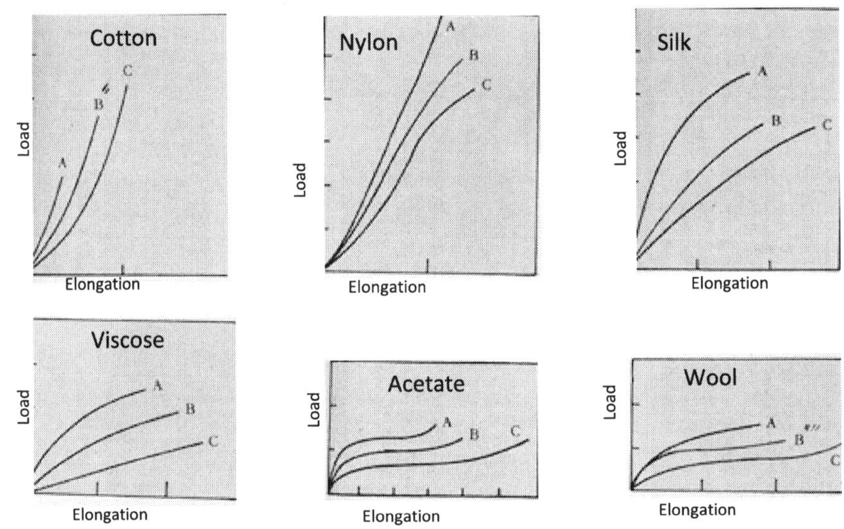

Figure 7.3 Effect of relative humidity on tensile strength of textile fibres A = 0% RH, B = 65% RH, C = 100%.

At 65% relative humidity, the more extensible fibres such as viscose rayon, cellulose acetate, silk, wool and, casein fibre have stress-strain curves which show an initial linear part, with stress proportional to strain followed by a gradual yielding of the fibre structure until extension increase considerably for only a small increase in applied load. This effect is more pronounced in wool.

Increase in relative humidity causes a very large decrease in the stress at which the increased rate of extension sets in. For some fi bres such as viscose and nylon, the yield region has disappeared at 100% relative humidity; the fibre being easily extended by the smallest applied stress. In this connection, it is interesting to note the peculiar behaviour of cellulose acetate. As the time of immersion of cellulose acetate fibre is increased, the yield stress also increases

The hydrophobic fibres such as polyester and polyamide show little difference in elastic behaviour, both in the wet and in the dry state. The hydrophilic fibres such as cotton, wool, silk, etc., exhibit marked changes in elastic behaviour when subjected to atmosphere of high humidity,

In vegetable fibres such as cotton, increase in strength is found with increase in relative humidity; whereas rayons and protein fibres decrease in strength with increase in relative humidity. The reason for this difference in behaviour lies in the fact that penetration of water molecules into the viscose rayon fibre, structure acts like a lubricant, weakening the lateral cohesion of the molecular chain. Thus, molecules slip past each other under tensile force and hence, register low breaking load. In case of cotton fibre, water molecule releases the initial stress in the long chain molecules attached to the crystalline regions and so increases fibre strength by a more uniform stress distribution. In other words, water molecules in the cotton fibre structure act like cement which is responsible for holding the chain-molecules intact. Thus, molecules do not slip past under tensile force and register high breaking load.

In case of cotton and other cellulosic fibres, extension-at-break increases with humidity. In case of polyamide and polyester, the increase in the extension is relatively small; the effect being more pronounced with viscose rayon and casein fibre. On the other hand, the initial Young's modulus decreases considerably as relative humidity increases.

Elastic recovery is very important for textile fibres if it has to be serviceable. This property is measured either as the ratio of extension recovered to the extension imposed or by the ratio of energy returned by the fibre to the energy absorbed. Elastic recovery for cotton, viscose rayon and, cellulose acetate decreases more rapidly as the extension is increased at 65% relative humidity than wool, silk and polyamide fibres. The increased relative humidity causes increase in the elastic recovery of wool, silk and polyamide fibres.

Elastic recovery of most of fi bres is affected by the moisture regain. Wet fi bres exhibit greater power of recovery than dry fi bres. At 65% relative humidity, glass fibre, nylon, saran, wool and fortison show an elastic recovery of more than 75%. Flax and silk show an elastic recovery between 50 and 75%. Cotton, Vinyon, Casein, and acetate rayon show an elastic recovery between 25 and 50%.

7.4 Practical significance of relative humidity

Textile manufacturing operations are conducted to a large extent in a humidified atmosphere. Under identical conditions, the following advantages are realised in processing:

1. Reduction in generation of static charges.
2. Reduction in the amount of dust and fly.
3. Allows for the retention of moisture already with the material.
4. Materials are more easily workable due to increased pliability.
5. Permits greater bodily comfort for personnel in cool weather.

8

Technical consequences and importance of some fibre properties

In this chapter, technical consequences of shape and dimensions of textile fibres and technical importance of some physical and chemical properties of textile fibres are discussed by considering both natural and manmade fibres. Further, some methods of altering the physical properties of the manmade fibres are given.

8.1 Technical consequences of shape and dimensions of textile fibres

One property common to all textile fibres is that they are all very long compared to their breadth. The material of which a fibre is constituted does not, by itself, determine that it will be a fibre. One would not think of a nylon molding or of a cellulose sheet as a fibre. It is, purely and simply, the shape that is characteristic of a fibre.

Length, fineness, cross-sectional shape, major heterogenetics, etc., are some of the important dimensions of fibres having decisive technical consequences.

The importance of the major structural parameters of fibres in technical processes and the properties of fabrics are summarised in Table 8.1.

Table 8.1 Technical consequences of shape and dimensions of textile fibres

Features	Technical properties affected
Fibre length	Drafting behaviour, yarrn strength, yarn hairiness, etc.
Fibre fineness	Spinning behaviour, limiting fineness of spinnable yarns, yarn evenness, distribution of fibres in blend, flexural and torsional rigidity, lustre of yarns and fabrics ease of uptake of dyes, tensile strength, etc.

Fibre cross-sectional shape	Flexural and torsional rigidity, lustre, density of packing of fibres, etc.
Major heterogeneities	Neps with immature fibres, lustre, flexural and torsional rigidity, handle, and bulk of fabric.
Axial form-crimp	Bulk of yarn, handle of fabrics

Since most of the fibre properties are interrelated, some of the technical properties affected by them are discussed here.

8.1.1 Length and fineness

Length of the fibre varies greatly within the sample, particularly with natural fibres (see Fig. 8.1). Fibre length determines smooth and successful working of many fibre processing machines. These machines are designed to operate on a narrow range of fibre length. A shorter mean length would result in a larger proportion of floating fibres and hence, in a poorer fibre control during drafting leading to more thick and thin places in the yarn. A greater proportion of short fibres would likewise lead to more thick and thin places and perhaps also to more neps in yarn. In combing, fibre length and variation in fibre length decides waste extracted. In roving and spinning, it determines the twist to be inserted and ultimately the breakage rate.

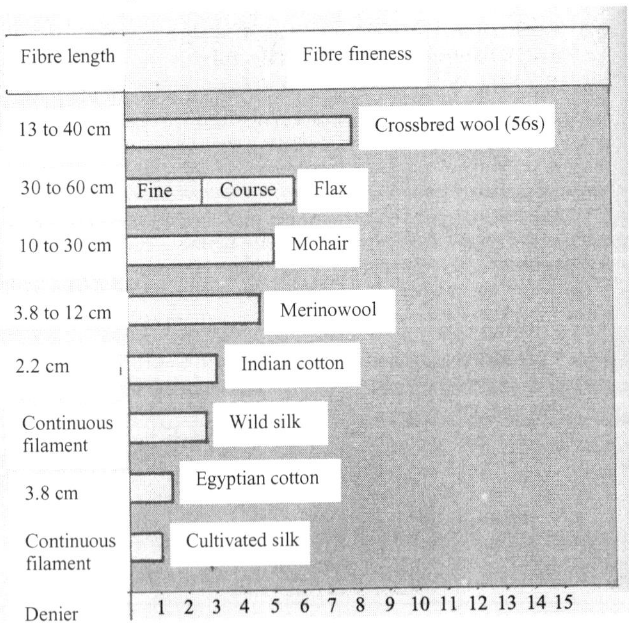

Figure 8.1 Length and fineness of natural fibres.

From the fabric point of view, fibre length has decisive effect on handle and feel. Short fibres projecting from surface produce hairy and warm handle and rough surface. Long fibres being fine produce comparatively fine surface.

For a given count, if one tries to spin finer count, number of fibres in yarn cross-section diminishes, the irregularity increases, and a point is reached where spinning any finer count further becomes impracticable. This is called limit of spinning.

Irregularity in the strand depends on the number of fibres in cross section which in turn depends on fineness.

Finer the fibre, finer will be the count that can be spun from the given fibre. Also, finer the fibre, greater is the length of fibre, and greater the surface area available for interfibre contact, more the cohesion with less twist and stronger the yarn.

For a given count, finer the fibre, the more uniform it is. Uniformity brings with it many desirable properties.

From the above discussion, it is clear that the length and fineness are inter-dependent factors having consequential effect on several mechanical properties and therefore, they influence the behaviour of fibre during processing. Fibre strength can indirectly affect the imperfection since, in weaker cotton, fibres are likely to break more during processing, and broken fibres would then behave as short fibres (*Textile Digest*, 1970, p. 61).

Resistance to twisting and bending is an important property that determines its utility. Resistance to twisting called torsional rigidity and resistance to bending called stiffness of a fibre are two factors to be considered before processing them. A fibre which is difficult to twist will probably be difficult to spin. It is interesting to note that the tendency of synthetic fibres to pill is the consequence of their high bending stiffness and their circular cross-section.

8.1.2 Fibre cross-sectional shape

Fibre cross-sections vary greatly from one fibre to another. Cross-sectional shape is important in lustre, bulk, body, texture, and hand or feel of a fabric. The shapes may be round, dogbone, triangular, lobal, bean shaped, flat, etc., as shown in Fig. 8.2.

The natural fibres derive their shape (a) from the way the cellulose is built up during plant growth, (b) the shape of the hair follicle and the formation of protein substances in animals, and (c) the shape of orifice through which the silk fibre is extruded.

The shape of manmade fibres is controlled by the spinneret and the spinning method. The size, shape, lustre, length and other properties of manmade fibres can be varied by changes in the production process.

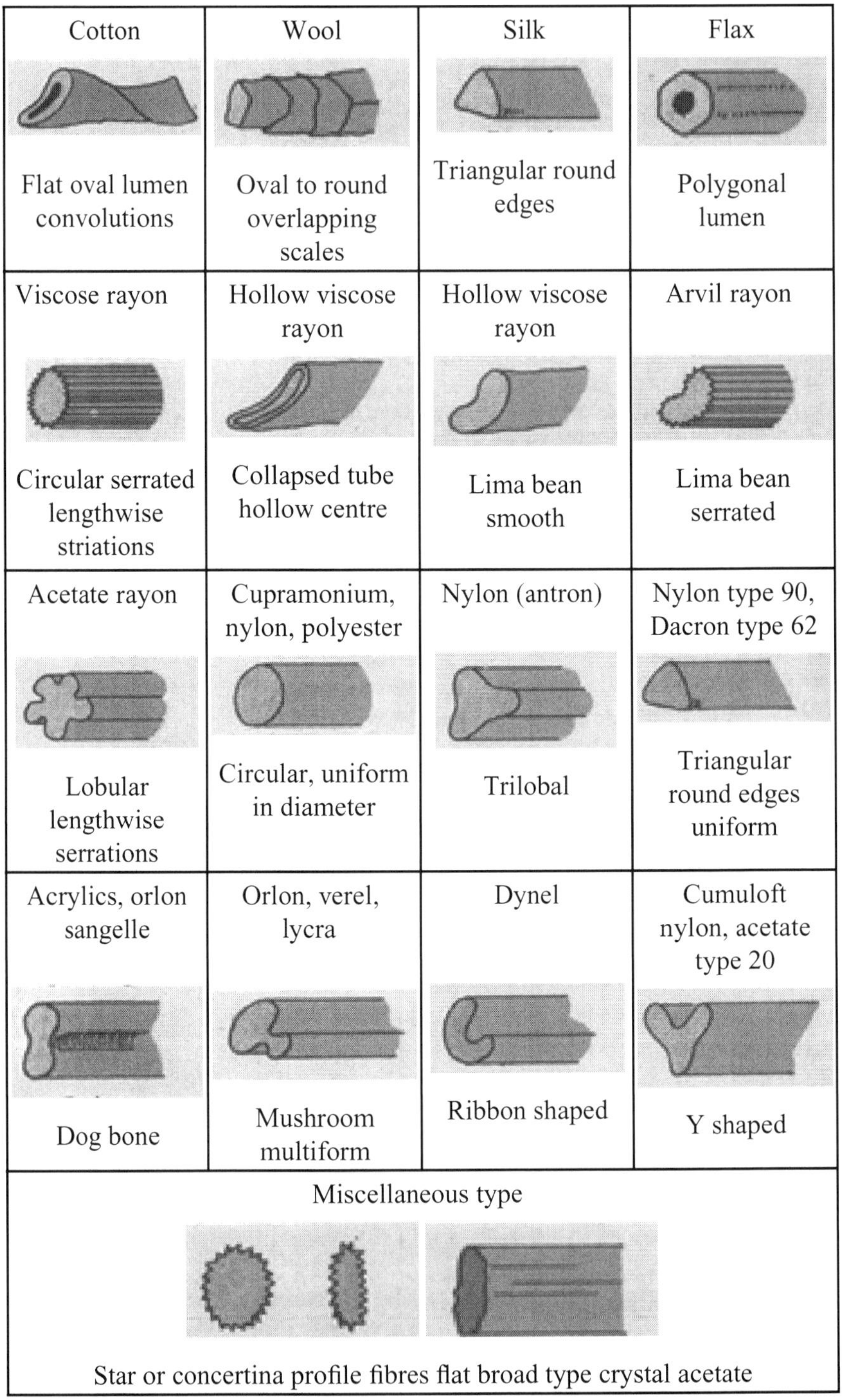

Figure 8.2 Typical cross-sectional shape and fibre contour.

From Fig. 8.2 it is seen that no two major natural fibres have the same fibre cross-sectional shape. Only in case of manmade fibres, some similarities in cross-sectional shape could be noticed; for example, nylon and dacron are almost circular in cross-section, other fibres such as vinyon and orlon have dumbell or dogbone sections.

Though rayons belong to one family of regenerated cellulosic fibre, they vary considerably in cross-sectional shape.

Cuprammonium filaments are almost circular, viscose rayon fibres are highly serrated; cellulose acetate fibres are lobed.

Generally, fibres that have been spun from a melt spinning technique have round sections; those that have been spun from wet or dry spinning have lobed or dogbone cross-sections. But if wet-spun fibres are greatly stretched during spinning, fi bre cross-sections will approximate to round (e.g., cuprammonium rayon and acrilan-acrylic fi bre).

Generally, flat sections have a high degree of lustre and excellent covering power, but have a harsh, unpleasant handle. An illustration of this defect of harsh handle being due to flatness of filament in alginate yarns.

Circular sections give a yarn, good handle and feel, but a poor covering power because the circular section offers the minimum sur face for a given volume.

It is interesting to note that amongst the natural fibres wool and the goat hair fibres have nearly circular sections and have very good handle, whereas cotton which is harsher, has a flat section.

It is found that the greater the surface area better will be the absorption of dyestuff, and it was probably for this reason that the serrated section of viscose rayon was developed; this type of section is produced by the presence of zinc salts in the coagulating bath.

The attractive handle of cellulose acetate is due to lobed cross-section. This cross-section is not very different from a tubular section but affords a better cover.

Trials have been conducted for determining the effect of non-circular section on nylon by spinning not through a circular jet, but through rectangular slots or triangular holes. An indispensable condition for experiments of this kind is that the melt, which is extruded, must be of a very high viscosity, mainly because if a relatively thin melt is extruded through holes of any shape, whatever the filaments will take up a circular section. But when a very viscous melt was extruded through rectangular slots (instead of the usual round holes), fibres of elliptical shape were obtained. There was neither marked difference in physical properties nor indeed in cover. In other words, elliptical and circular section nylons did not show much difference, but some yarn when spun with triangular section filaments did show significant differences.

There is no doubt that the triangular section increases the covering power. It has also influence on the handle of fabric made from it. Real silk is composed of filaments which approximate to triangles in cross sectional shape. Another point observed was that the fabrics from filaments of triangular section were lustrous.

8.1.3 Major heterogeneities

Fibre maturity is one of the important characteristics of cotton fibre. The maturity of cotton fibres varies not only between fibres of different samples but also between fibres of the same seed. The presence of excessive percentage of immature fibres in a sample is undesirable as this causes excessive waste losses in processing and lowering of the yarn appearance grade due to formation of neps (*Textile Digest*, 1965, October/December No. 4, pp. 201-205). Neps appear as ugly specks in white cloth, and as unevenly dyed spots in coloured fabrics. Neps also tend to lower yarn strength and increase end breakages during spinning. (Neps are small knot-like aggregates of highly entangled cotton fibres, usually not longer than a common pin head, which are difficult to remove) The chief cause of neppiness is a larger number of immature or dead fibres in cotton lint. These fibres tend to curl up when subjected to beating or rubbing action and in the process of curling up they get entangled with the adjoining fibres forming neps.

8.1.4 Crimp

Fibre crimp is one of the important factors which have direct technical consequence on the yarn and the fabric manufactured. Fibre crimp refers to the waves, bends, twists, coils or curls along the length of the fibre. Fibre crimp is measured by the number of arcs per unit length. Fibre crimp increases cohesiveness, resiliency, resistance to abrasion, stretch, bulk and warmth. Crimp increases absorbency and skin contact comfort but reduces lustre. A fibre may have one or three kinds of crimps (a) natural or inherent crimp which occurs in cotton and wool (b) mechanical crimp which is imparted to fibres by a process called Texturising and (c) the latent crimp which exists in an undeveloped state in bicomponent manmade fibres.

8.2 Technical importance of some physical and chemical properties of textile fibres

8.2.1 Physical properties

The end-use suitability of a fibre depends upon the technical importance of its properties. A certain fibre, used in climbing rope, may not find its utility in a parachute fabric or in beltings and tyre cord and vice-versa. Therefore, an understanding of the properties of fibre and its technical significance paves way for better utilisation of fibres for a specific end-use. Following are some of the physical properties of technical importance:

- Tensile strength
- Breaking elongation
- Elastic recovery
- Work of rupture
- Initial modulus
- Tensile Strength
- Stiffness
- Specific gravity
- Effect of moisture
- Thermal properties
- Effect of sunlight

Tensile strength: This is the breaking strength of any material expressed as force per unit cross-sectional area, e.g., as pounds per square inch (psi) or kilogrammes per square metre (ksm). In these terms we describe the ability of any material to resist breakage when it is under tension.

Two fibres with identical tenacities may have different tensile strengths. It is the density factor which divides them. If densities are of different, the cross-sectional areas too will be different. Tenacites of common fibres are shown in Fig. 8.3. Strength or tenacity, thus, gives a measure of resistance to steady forces. It will, thus, be correct quantity to consider when a specimen is subjected to a steady pull, e.g., rope for lifting heavy weights.

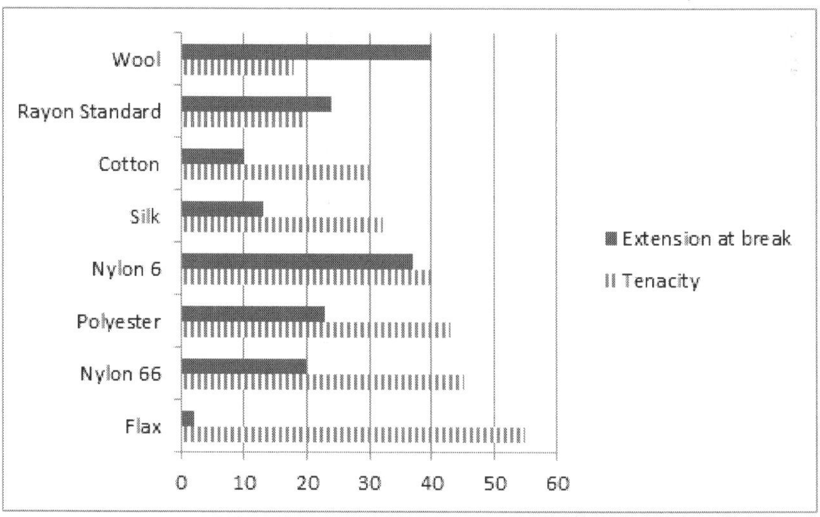

Figure 8.3 Tenacity and extension at break of some fibres.

Breaking elongation: When a fi bre is subjected to a force it will stretch to certain degree. This stretching is described as elongation of the fi bre in terms of percentage of the fibre and original length. It can be measured either as an elongation under a certain load or as the elongation reached when fibre breaks. It is, thus, important when the specimen is subjected to stretching as in the case of the neck of the garment being pulled over the head.

Elastic recovery: When a fi bre is stretched by a small amount, it may exhibit almost perfect elasticity, i.e., it will return to its original state when released. Elastic recovery in this case is 100 percent. If, however, the fibre is subjected to a higher stretch, it may react in much core complex way. Some permanent deformation may take place so that when fibre is released it may return to an elongated form. It recovers from some of its elongation but not all. This behaviour of a fibre can be denoted by describing its elastic recovery at certain elongation. Thus, when the fibre returns completely to its original length, after, say, 2 percent elongation we can say that elastic recovery is 100 percent at 2 percent extension. Elastic recovery of any fibre contributes to dimensional stability in a fabric. Elastic recoveries of some common fibres are shown in Fig. 8.4.

Work of rupture: A fi bre is required to withstand sudden shocks in some applications like the climbing rope, a rope meant for lifting a heavy load, rope or twine used in parachute, etc. A shock contains a given amount of energy and the fibre must be able to absorb the energy without rupture. Thus, work of rupture is a measure of the toughness of the fibre.

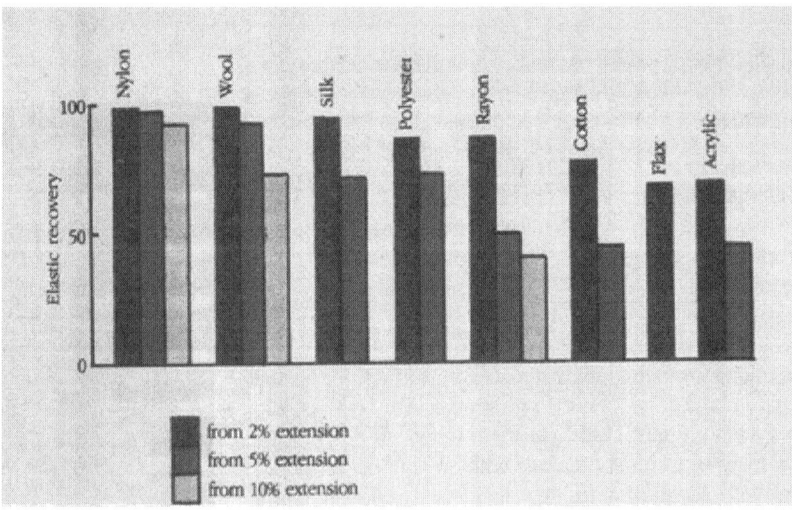

Figure 8.4 Elastic recovery of some common fibres at 65% RH and 20°C.

Initial modulus: When a fi bre is loaded, it extends. Resistance to small initial extension is measured as initial modulus. When the fibre has high resistance to stretching, it will have high modulus. A low modulus fibre requires little force to stretch it.

A fibre which will be most satisfactory for textile purposes is the one that does not stretch much for small loads. The main advantage of this is that even if the winding tension varies slightly from spindle to spindle, the yarn will not suffer seriously different degrees of stretch from spindle to spindle: as it would if it were susceptible to high degree of stretch from a very small load. The fibre, that is best for soft and comfortable wear, is likely to be the best for industrial uses such as belts and ropes, where high strength and good recovery are essential.

Stiffness: Stiffness is the ability of a fibre to carry a load without deformation. This is based on modulus of elasticity, expressed in a unit of gpd.

The fibres which are wiry are less adopted for textile purposes. Therefore, fibre should have the property of flexibility combined with its ability to carry load without deformation. This puts these fibres in the major circle of utility fibres. A high degree of bending stiffness is desirable in carpets. It is found that bending stiffness of a flattened cross-section is less than that of a round cross-section of equal cross-sectional area.

Specific gravity: It is a relative measure of the density of a material. It is deemed as the ratio of mass of material to the mass of an equal volume of water at 4 °C.

This is an important characteristic of any material. It affects the drape of a fabric. Fibre of low density tends to give bulkier yarn for a given linear density.

Effect of moisture: The ability of fi bre to absorb moisture, will also affect the processing and fi nishing of yarns and fabrics. A high degree of moisture absorption indicates the ability to absorb perspiration which is essential for clothing fi bres. Dyes readily penetrate into the fi bres which have the ability to absorb moisture.

Synthetic fibres having low moisture absorption are easily washed and dried compared to other fibre. They tend to accumulate charges of static electricity much more readily than moisture absorbing fibres. Mechanical properties of textile fibres are also affected by the moisture.

Electrical properties: Electrical properties of a fi bre determine the degree to which static charges will accumulate on a particular textile during processing and in use. Static charges may be produced due to friction between yarns and fabric and the surfaces they meet on processing machinery.

Static electricity does more harm than good. It often causes serious processing troubles resulting from severe shock, fire risk, and choking of machine parts. It also attracts dust, fluffs, causing damage to the machine, and reducing the production and increasing the waste during processing.

Production of static electricity is greatly influenced by moisture effect on fibres. A damp fibre will conduct electricity away as soon as it is formed, so that pools of static charges do not collect on fibre. Following are some of the precautions to reduce generation of static charges during processing:

- By earthing the machine,
- By using antistatic dressing agents,
- By using RASE (Radio Active Static Eliminator), and
- By increasing the humidity.

Electrical properties of fibre are often considered very important if they are to be used as insulating material in the electrical industry.

Thermal properties: The response of fi bres to differences of temperature is of great technical value and importance, particularly within the range of temperatures that are met in practical use as well as in wet processing.

In the presence of air, most fibres will burn. The readiness with which these catch fire and support combustion is of immense practical importance. Many accidents are caused by clothing catching fire. Hence, the increasing need for a nonflammable textile is felt.

Many of the new synthetic fibres are the thermoplastic, i.e., they will soften while they are heated. The temperature at which they soften largely determines their practical usefulness in textile field, that is, in the manufacture of textured yarns.

Effect of sunlight: Powerful radiations of sunlight affect almost all fibres. Many will deteriorate rapidly, lose their tensile strength and get decomposed besides changing their colour. A fibre which resists photo degradation is useful for door hangings and curtains.

In a study on the effect of sunlight on curtains, following rating has been given for various fibres in order of increasing rate of loss of strength on exposure to sunlight:

1. Fibre-glass
2. Acrylics
3. Polyesters
4. Cellulose diacetate and macerate
5. Nylon 6 and nylon 66

6. Linen
7. Cotton
8. Wool
9. Silk

One must remember that these fibres have very different initial strengths, which must be taken into account as well as the rate of strength loss, when trying to decide on the best blow from the point-of view of light stability.

8.2.2 Chemical properties

Modem techniques of processing fibres, yarns and fabrics involve use of lot of chemicals. Alkaline scouring agents, bleaching and dyeing assistants, detergents, and other chemicals are used in preparing finished textiles. The fibre itself must withstand these substances without suffering hamlfull effects viz., decomposition, degradation, deterioration, purity, conversion, hydrolysis, oxidation and reduction.

Following are some of the important chemical properties of technical importance:

1. Effect of acids and alkalie,
2. Effect of organic solvents, and
3. Effect of micro-organisms and insects.

8.2.2.1 Effect of acids and alkalies

Effect of acids and alkalies must be known to technologists since many textiles are commonly subjected to acid and alkali solutions. A better understanding of the fibre nature and its behaviour with concentrated and dilute acids and alkalies have led to the introduction of different types of finishes on fibre, yarn and fabric for instance, mercerisation introducing stretch to yarn, varieties of finish and prints, etc.

Introduction of certain stain removers are based on the understanding of the basic behaviour of acids and alkalies on fibres.

8.2.2.2 Effects of organic solvents

The processes and materials used in laundering the fabrics are mainly responsible for retaining the strength and serviceability of the material. In order to take care of the materials, knowledge of effect of organic solvents on a textile fibre is of paramount importance.

Certain types of clothing fibres, such as wool and silk, require dry cleaning treatment. If these fibres are subjected to regular laundering treatment where concentrated alkaline solution in water is used, it may result in damage for the fibres and reduction in its elegance, warmth giving property, etc. Therefore, fibre itself must obviously resist these solvents to assure stability of structure. Introduction of dry cleaning has given solvent-resistance a great significance in textiles.

8.2.2.3 Effect of micro-organisms and insects

Mildew and bacteria affect cellulose. They decompose it and make use of the degradation product as food. Textile products stored in warehouses are always affected by mildews. This can discolour and weaken the fibre to the point where they become useless.

Certain cellulose base and protein base fibres are likely to be attacked by insects and are enjoyed as a privileged food by other insects. Synthetic fibres are much more resistant to biological attack.

Some of the physical and chemical properties discussed above must be taken as the basic characteristics of a fibre which has more technical importance than their existence.

8.3 Some methods of altering the physical properties of manmade fibres

Manmade fibres can have their physical properties altered during manufacture. There are two main approaches to alter the physical properties of fibres, namely,

a. Physical modification, and

b. Chemical modification.

Physical modification can be done by the following methods:

- Altering the shape of the spinneret holes,
- Altering the take up velocity,
- Adding delustring agents or pigments,
- Altering the draw ratio,
- Changing the composition and condition during spinning, and
- Changing the polymeric components.

Chemical modification includes the followings:

- Cross-linking
- Copolymerization

8.3.1 Physical modification

8.3.1.1 Altering the shape of the spinneret hole:

The cross-sectional shape of the filament may be varied by extruding spinning solution through spinneret holes of the desired shape. Modification of filament cross-sectional shape is important as it can cause profound change in the characteristics of yarns and fabrics. Circular cross-section filament has poor covering power. Though flat filament made by extrusion of polymer through slit orifices (instead of circular) have got good covering power, it tends to possess harsh handle. Triangular cross-section is derived from a particular shape of extrusion holes. It offers great relative covering power and possesses better lustre than circular cross-section filaments. Relative covering power of a fibre is the amount of fibre required to cover the same area.

The fibre which has the lowest specific gravity not only results in light weight fabrics from the standpoint of comfort and ease of handling but also provides extra cover per kilogram of material. Table 8.2 shows the comparable specific gravity and cover for different fibres as compared to Proplon–the polypropylene fibre.

Table 8.2 Relative covering power of textile fibres

Fibre type	Relative covering power (kg)	Specific gravity
Proplon	1.00	0.91
Nylon	1.26	1.14
Polyester	1.52	1.38
Acrylic	1.30	1.17
Viscose rayon	1.70	1.52
Wool	1.43	1.33
Cotton	1.65	1.50

8.3.1.2 Altering the rate of delivering of the spinning solution through meter pump:

The rate at which the spinning solution is pumped to the spinneret determines the denier of the filament spun. Metering pump ensures constant and regular supply of the spinning solution to filter and spinning jet at the exact rate required for the denier being spun. Therefore, alteration in speed of metering pump, rate of delivery relative to the rate of draw-off take up enables the denier of the filament to be changed.

8.3.1.3 Altering the take-up velocity:

In the manufacture of PET (poly-ethylene terephthalate) fibres, when the take-up velocity is increased from 1000 to 6000 metres/minute, the orientation of the molecules in as-spun filaments keep on increasing leading to changes in the mechanical properties. For example, upto the take-up speed of 1000 metres/minute, the orientation is found to be very poor. Such PET yarns spun are designated as low-oriented yarns (LOY). In the region of take-up velocity from 1000 to 4000 meters/minute, the as-spun filaments of PET show a comparatively higher molecular orientation but no crystallinity. These types of yarns are designated as partially-oriented yarns (POY). With the velocity above 4000 m/min, the highly oriented and crystallized filaments can be produced and these are called as fully-oriented yarns (FOY).

From the above discussion, it is seen that the molecular orientation and degree of crystallization increases with an increase in take-up speed. Hence, we can obtain various fibres from unoriented yarns (UOY) to FOY by altering the take-up velocity.

8.3.1.4 Adding delustring agents or pigments:

Dull finish can be imparted to the filaments by adding delustering agent or pigments. The most common delustering agent is Titanium Dioxide and is used to impart dullness to viscose rayon, nylons and PET filaments.

8.3.1.5 Altering the draw ratio:

Normally, the stretching is imparted to the filaments by running the two godet wheels at different speeds. Stretching is done mainly to orient the molecules of fibre along the fibre axis. The degree of orientation depends upon the amount of stretch to which filament is subjected. Tensile properties can be controlled by controlling the degree of stretch. As drawing continues, more and more molecules are brought to the state where they pack to form ordered regions. The molecules are held together as a result of the cohesive force.

8.3.1.6 Altering the composition and conditions during spinning:

The composition of the polyester being extruded also affects the physical and chemical properties of the fibres.

In viscose rayon manufacture, sodium sulphate precipitates the sodium cellulose xanthate from viscose solution into the form of filaments and sulphuric acid converts it into cellulose. Glucose gives pliability and softness to the yarn probably because it increases the viscosity of coagulating bath and thus retards a little, chemical change from sodium cellulose xanthate to cellulose. Zinc sulphate is responsible for added strength and also for serrated cross-section. If it is omitted, filaments are round or oval in section and do not dye so well as when they are serrated or indented.

8.3.2 Chemical modification

Cross-linking is one of the major chemical modifications carried out on manmade fibres for altering their physical properties. For example, long chain molecules of polyamide may be linked by reacting them with isocyanate group or any chemical having reactive group at either ends. Similarly, cross-linked rayon fibres are obtained from intermolecular cross-linking of cellulose by mixing high molecular substances with viscose. It is characterised by low water and alkali swelling and is capable of withstanding washing. It is chiefly used for manufacturing quilts, carpets, curtains, and upholstery fabrics.

8.3.2.1 Copolymers

Copolymer is the polymer formed from two or more different monomers. There is no definite regularity in the order of monomers which makes up a copolymer: For example,

If monomer is A or B or C

then, polymer may be – A – A – A –A – A – A

or

– B – B – B – B – B – B –

or

– C – C – C – C – C – C –

copolymer will be

A – B – A – A – A – B –

or

A – C – A – C – C – A –

or

C – A – A – C – A – A – A – B –

Depending on the polymerization condition, copolymers are subdivided into (i) Random, (ii) Graft and, (iii) Block copolymers.

Random copolymers: In this, two monomers A and B are presented simultaneously in the reactor, giving rise to a macro molecular chain

(A – A – A – A – B – B – A – B – A)

Graft copolymers: Here, the polymer (of monomer A) is prepared, and one more monomer is grafted onto this polymer. The fi nal product consists of polymeric backbone with one or more side branches such as

Crafting of polymer on the sides of the polyamide is a chemical technique yielding substantial results. For example, by the use of grafting technology, it may be possible to impart oil and water repellency, flame retardance, improved dyeability and ion-exchange properties to rayon fibres. Although a laboratory curiosity, graft copolymerisation of various polymerizable monomers into rayon fibres has been known for quite sometime, it is only recently the formidable practical limitations have been overcome and commercial production of graft fibres on a limited scale has been reported.

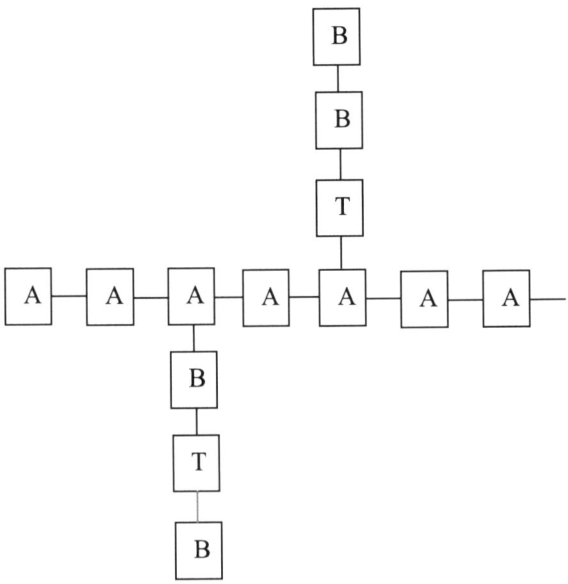

Block copolymers: The block copolymer is a linear chain with a sequence of different polymer segments, i.e., (– A – A – A – A – BB – B). In this case, one monomer (A) is polymerized, and another monomer (B) is polymerised onto the living ends of the polymeric chains.

9

Identification of textile fibre properties

In the early days, when almost the textile materials for domestic use were made from cotton, wool and linen, it was comparatively easy to identify fibres by using burning tests, staining tests and microscopic examination. But in the recent times, due to the introduction of many artificial fibres, fibre identification has become much more complicated. The increasing of use fibre blends has aggravated the complication of fibre identification.

At present, there are atleast fourteen different fundamental types of fibres, namely, cotton, wool, silk, linen jute, cellulose esters, regenerated cellulose fibres, polyamide, polyester, polypropylene, acrylic, etc. Many of these fibres are available in modified forms, e.g., acrylic modacrylics, nylon 6, nylon 66, bicomponent fibres, etc. It is, thus, not possible to describe here a single identification test, which could indicate with certainty, the type of fibre present in any textile material. It will be sufficient to state that identification requires the application of several tests based on the following:

1. Microscopic examination of the fibres and also its cross-section,
2. Behaviour and solubility in various acids and alkalies and organic solvents,
3. Behaviour of the fibre under heat and flame, also melting point of the fibre,
4. Reaction to various colour staining solutions including Shirlastain,
5. Moisture regain characteristics,
6. Refractive index value,
7. Specific gravity value,
8. Feeling test, and
9. Breaking test.

No single test, other than microscopic test, can be depended upon to give immediate recognition of the fibre. Limitations are encountered in ail the tests. In practice, identification tests are used in combination.

Certain tests cannot be made freely or easily in the retail store at the time of purchase. At that time, consumer must use non-technical tests, viz., No.3 and 9, mentioned in the list. These are inadequate, yet they convey a certain amount of information if they have been used frequently engough to ensure correct observations.

9.1 Microscopic tests

9.1.1 Morphology and structure of fibres

Morphology refers to form and structure of a substance. Each textile fibre has its own distinct structural markings. Under a microscope with a mangification of atleast hundred, it is possible to make clearly an absolute identification of fibres. The differences in morphological structure determine the various characteristics of the different fibres and explain why some fibres are to be preferred to others for certain uses.

9.1.1.1 Cotton

A matured cotton ftbre under microscope appears as a flattened spirally twisted tube with characteristic central canal or lumen extending throughout its length. The natural twist of the cotton fibre is called convolutions. The convolutions range from 150 to 300 per fibre, depending upon the type of fibre. Convolutions for sea island type are 14 per mm, for Egyptian cotton 12 per mm, for American cotton 8 per mm, and for Indian varieties 6 per mm.

Convolutions are almost absent in immature cotton fibres. The base of the fibre is broad owing to removal of the fibres from the seeds during ginning while the tip is tapered and closed.

Radial cross-sections are often bean-shaped configurations ranging from U shape to a circular (See Fig. 9.1) depending upon the degree of maturity. Fibres from different varieties of plant vary considerably in length and fineness. Short staple fibres for instance are about 12.7 mm long, long staple types may exceed 50.8 mm. The range of fibre width is about 12 to 26 microns. In the longitudinal cross-section, depending upon the width of the lumen (L) and the wall width (W) of a fibre, maturity characteristics are identified (Figs. 9.1 and 9.2). According to Indian Standards Institution, (IS-1611-1960) depending on the ratio of lumen width (L) to wall thickness (W), all fibres are classified into following three categories:

$L/W < 1$ for matured fibres

$L/W < 2$ for half-matured fibres

L/W ≥ 2 for immature or dead fibres

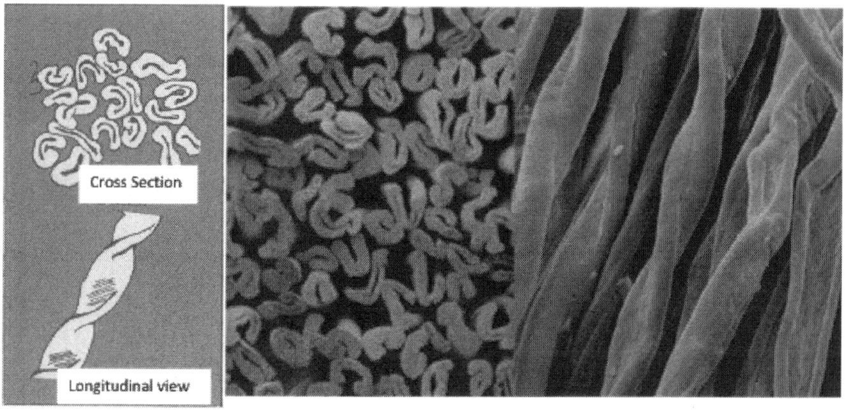

Figure 9.1 Microscopic view of cotton fibre.

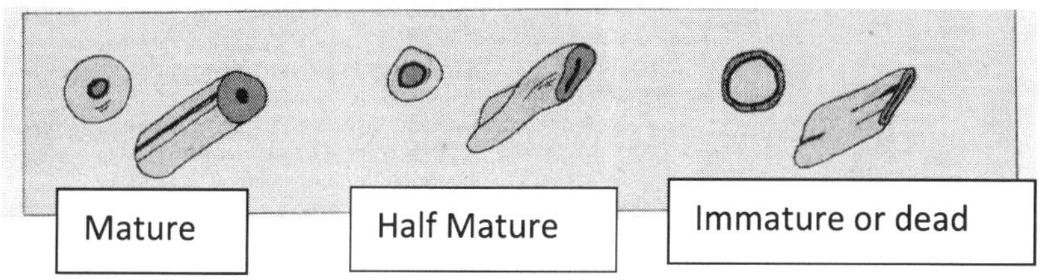

Figure 9.2 Maturity characteristics of cotton.

The microscopical appearance of mercerised cotton varies depending upon the extent of mercerisation. Fully mercerised fibres are characterized by lack of structural details. The fibres are smooth surfaced and cylnidrical and show few convolutions owing to the swelling which occurs during mercerisation. The lumen is reduced to a narrow line. In cross-section, the fibres appear round to oval with a small point indicating the position of the lumen.

9.1.1.2 Wool

In common with animal hairs, the surface of sheep wool fibres is made up of scales. The size of the scale varies. It may be very small or broad and large. When counted, as many as 2000 scales are found in one inch of fine wool while coarse wool may have as few as 700 per inch. Fine wool fibres do not have as clear and distinct scales as coarse wool, but they can be indentified under high magnification.

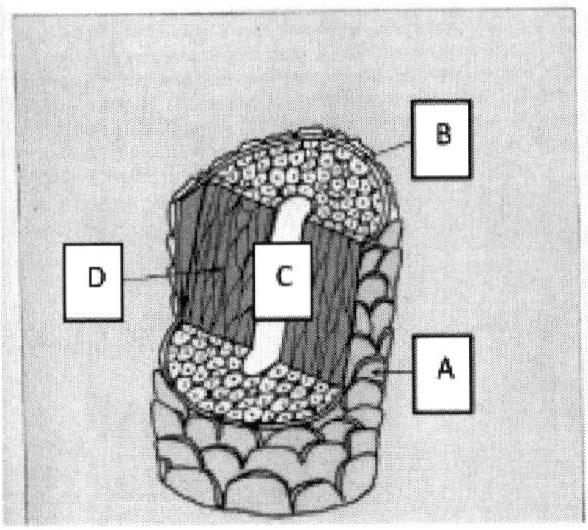

Figure 9.3 Sectional view of medulated wool.

(A. Cuticular scales B. Cortical cells C. Medula D. Fibrillar structure)

From the Fig. 9.3, it is seen that the scales which overlap the free ends of the scale project outwards and point towards the tip of the hair, thereby, giving the outline of the fibre a barbed appearance.

The major portion of the fibre is the cortex composed of cortical cells and extends towards the centre from the cuticle layer. Cortical cells are long and spindle shaped cells and provide strength and elasticity to the fibre. Cells range from 80 to 110 microns in length and 2 to 5 microns in width and 1.2 to 2.6 microns in thickness. The cortex accounts for approximately 90 per cent of the fibre mass. The faint longitudinal striations observed in the wool fibre are attributed to the cortical cell.

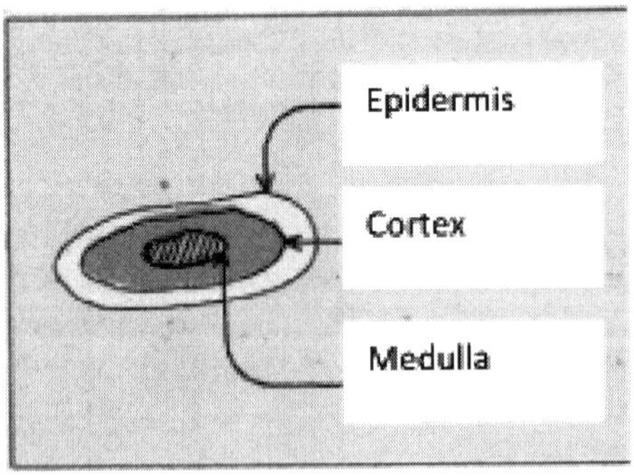

Figure 9.4 Position of medulla in wool.

The innermost layer is the medulla (Fig. 9.4), which consists of pith of oval shaped cells, is usually air filled. The medullation is discernible only on coarse and medium quality wool. The medulla appears from a narrow to a broad line or from a continuous to an interrupted line.

The medulla is built up of many superimposed cells, of various shapes, often polygonal, forming a honey comb-like structure.

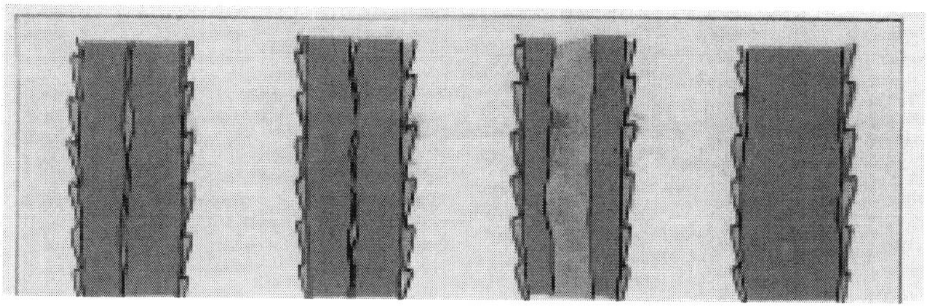

Figure 9.5 Medullation of wool fibres.

The diameter of the cells varies from 1 to 7 microns. Various porous channels pass through the medulla cells which are normally filled with air. The shape and size of the medulla vary greatly. The medulla might consist of a single chain of cells or of several series arranged side by side, it varies from 10 to 90 per cent of the whole fibre. Accordingly, they are called Hairy, Hetero and Kemp type (Fig. 9.5). Very fi ne grade of wool does not contain medullation and are called true fi bres.

The presence of medullated fibres in any wool is detrimental to quality from the standpoint of the manufacture. The spinning properties of medullated wool will be inferior and in piece-dyed fabrics. They produce a neppy effect by dyeing a lighter shade. The medulla is in no way necessary for the growth of the fibre; its main function is to increase the protective properties of the fibre by adding internal air spaces.

The wool fibre cross-section is nearly circular but most wool fibres tend to be slightly elliptical or somewhat oval in shape, as shown in Fig. 9.6.

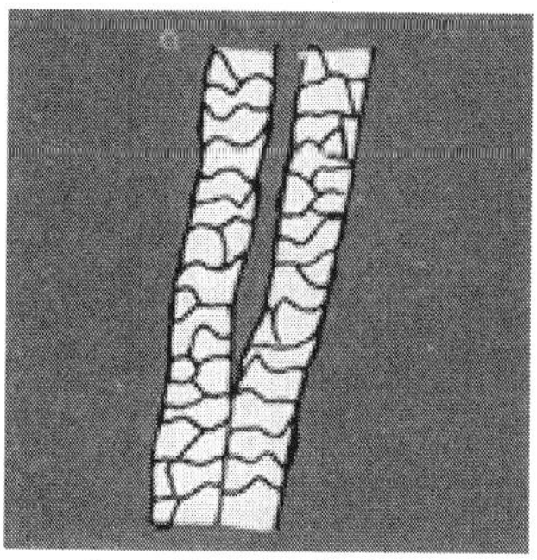

Longitudinal view

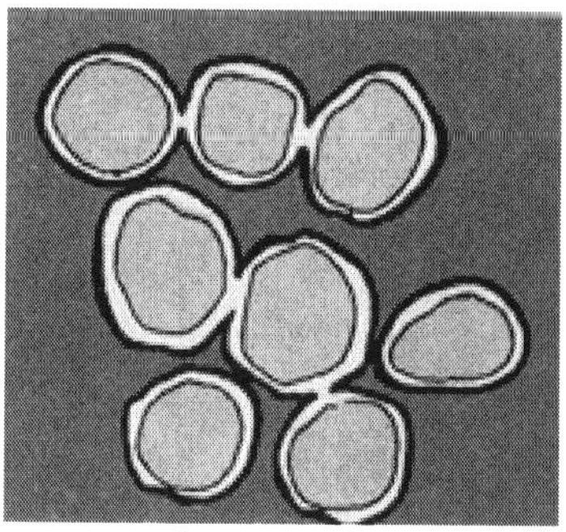

Cross section

Figure 9.6 Microscopic view of wool.

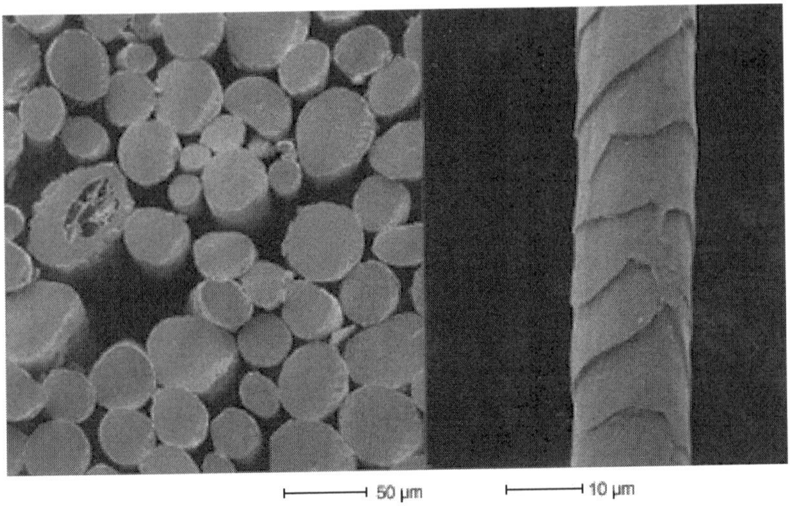

The width of the wool fibre varies considerably. Fine fibres, such as Merino, possess an average width of about 15–17 microns. Medium wool averages 24 –34 microns and coarse wool is about 40 microns. Kemp fibres, those which are exceptionally stiff and coarse, are about 70 microns in width.

With wool fibre, a built in waviness, called crimp, is distinctly visible to the naked eye. The number of crimps in very fine wool range from 55.88–76.2 per centimetre and 35.56–55.88 crimps per centimetre. Coarse wool contains less than 25.4, and very coarse wool has two to three crimps per centimetre, sometimes nil.

In general, in fine wool more number of crimps per centimetre are observed than in coarse wool fibres.

Fine and medium wool fibres tend to have more lustre than coarse fibres.

9.1.1.3 Silk

Silk filaments are very fine and long. They mostly measure from about 360 metres to 1238 metres in length and can be as long as 3000 metres.

Cross-sectional view of silk shows triangular shaped fibres with no markings. Two filaments usually lie with two flat sides, each of which is called Brin (Fig. 9.7), held together by sericin, a gummy substance that gives the Bave (cultivated cocoon) a rather uneven surface. As the sericin is removed by hot water (degumming), the two filaments appear as a smooth transparent rod in longitudinal view (Fig. 9.8). Wild silk tends to be quite uneven and is somewhat dark and may have longitudinal striations.

The width of the silk fibre ranges from 9 to 11 microns. The fibres are quite smooth. They have a high natural lustre or sheen and are of white to cream in colour. Wild silk is uneven with slightly less lustre and tan to light brown in colour.

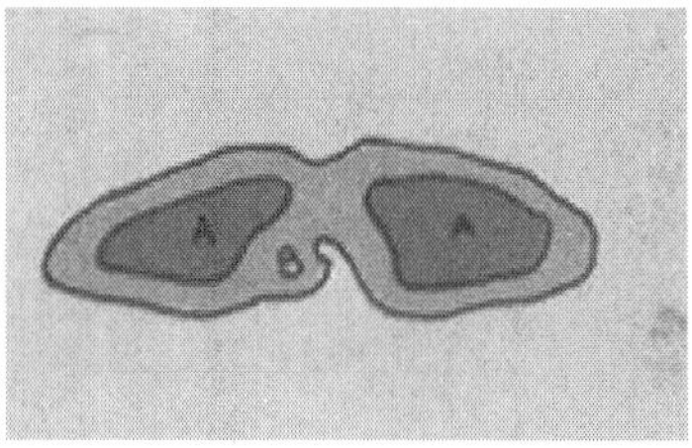

Figure 9.7 Cross-section of raw silk filament showing two separate fibres "A" cements together side by side with silk gum or sericin "B"

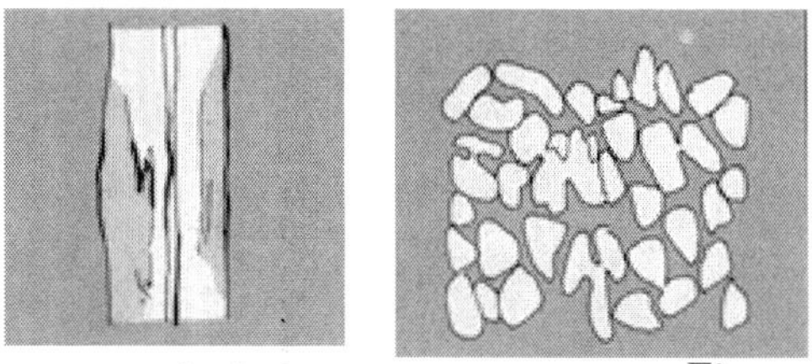

Figure 9.8 Microscopic view of silk fibres.

9.1.1.4 Jute

Individual jute fibre cells range from 1.52 to 5.1 mm in length and their diameter ranges from 20 to 25 microns. The surface of the cell is generally smooth with only occasional nodes or cross- markings. The cross-section of the fibre is polygonal in shape with sharply defined angles (Fig. 9.9). Cell walls are thick. The lumen is large and the cross-section is oval. Longitudinally, the lumen is irregular with occasional sharp constructions. Towards the end of the cell, the lumen thickens, resulting in the cell wall becoming very thin and the cell ends tapering off. The cell is surrounded by a thin medium layer of lignin

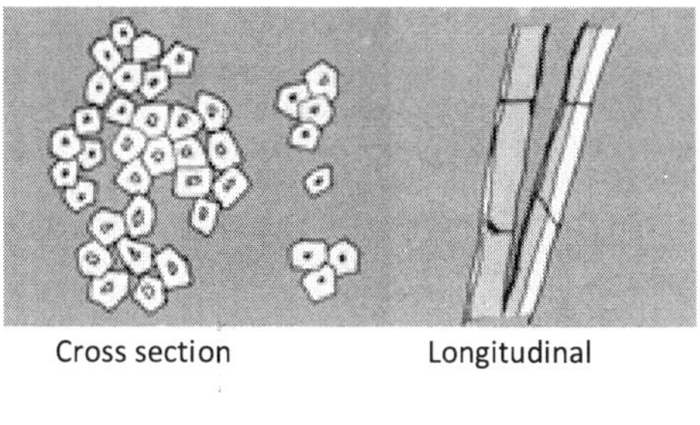

Figure 9.9 Microscopic view of jute fibres.

9.1.1.5 Flax

The flax fibre is composed of fibrils or bundles of fibre cells held together by a bonding or a gummy substance. Under the microscope (Fig. 9.10), the longitudinal view of the fibre shows the width to be quite irregular. The central canal or lumen casts a shadow giving a slightly darker effect down the centre. There are no convolutions as in cotton, but longitudinal lines may be seen. The points at which the fibre width changes are marked by swellings and irregular joint formations called nodes, which appear similar to the joints on bamboo.

The cross-sectional view shows the lumen quite clearly, the thick outer wall and a round to polygonal shape. Immature flax may be oval in shape and usually has a larger lumen than a mature fibre. Width of the flax fibre ranges from 15 to 18 microns. Flax possesses a high degree of natural lustre. The natural colour of flax varies from light ivory to dark tan or grey.

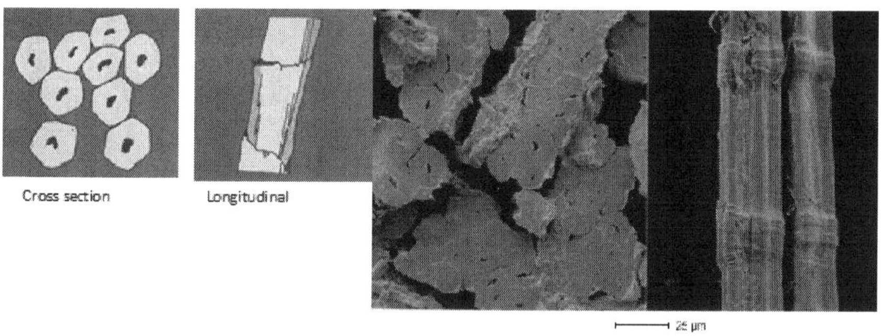

Figure 9.10 Microscopic view of flax fibres.

9.1.1.6 Viscose rayon

The longitudinal appearance of regular viscose rayon exhibits uniform diameter and interior parallel lines. If the fibre is delustred, it shows a grainy and pitted appearance. Bright fibre will appear relatively transparent.

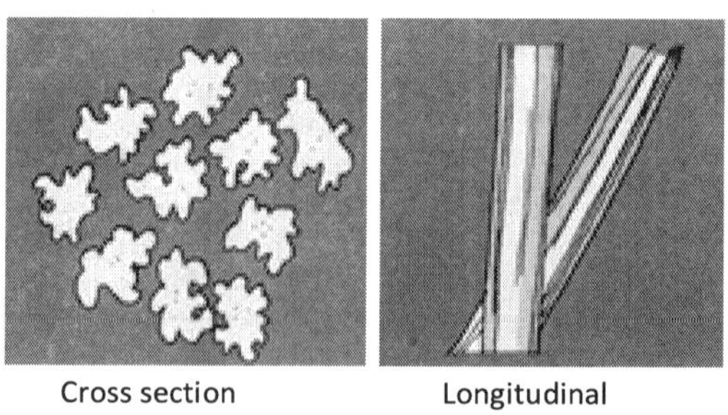

Figure 9.11 Microscopic view of viscose rayon fibres.

The cross-section of the fibre shows highly irregular or serrated edges (Fig. 9.11). The presence of delusterants is obvious by a spotted effect, white bright fibre, which appears crystal clear. The width of the filament ranges from 8 to 43 microns depending upon the denier of the fibre. The striations, seen in the longitudinal view, are the result of light reflection by the irregular surface contour.

High tenacity viscose rayon is similar to regular viscose when viewed under magnification, except that it may have less irregular contour in cross-section and, thus, show fewer striations in the longitudinal view.

9.1.1.7 Cuprammonium rayon

The filament appears uniform in width and these surfaces are without any marking under the longitudinal view. The cross-section of cuprammonium rayon is round or oval (Fig. 9.12). It appears relatively clear and there are no irregularities in the contour.

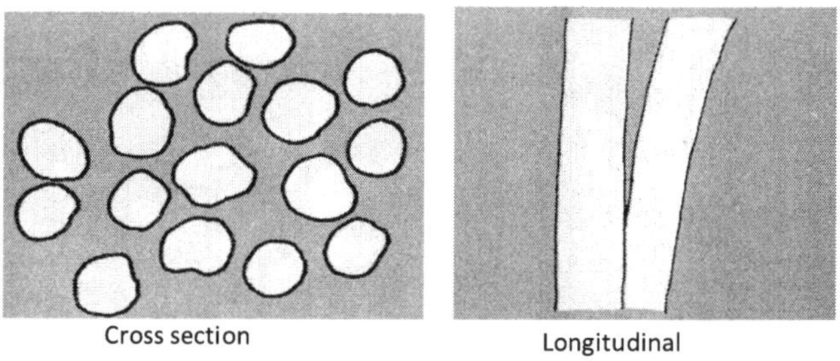

Figure 9.12 Microscopic view of cuprammonium rayon fibres.

The average fineness of the cuprammonium filament is 1.25 denier, but these filaments have been made as fme as 0.40 denier. The variation in the denier of the cuprammonium filament is on the average of about 5%.

9.1.1.8 Acetate rayon

The cross-section of acetate rayon fibre is somewhat like a clover leaf (Fig. 9.13). The individual lobs, which number two to four, are round and smooth, and are easily distinguished from the more numerous shallow indentations which occur in viscose rayon fibres. The speckled effect is noticed in dull acetate, but bright fibres appear clear. In the longitudinal view, the clefts between the lobs show up as two or three pronounced striations. Acetate fibres are less lustrous compared to viscose and cuprammonium rayon. The width of the fibre ranges from 11 to 46 microns.

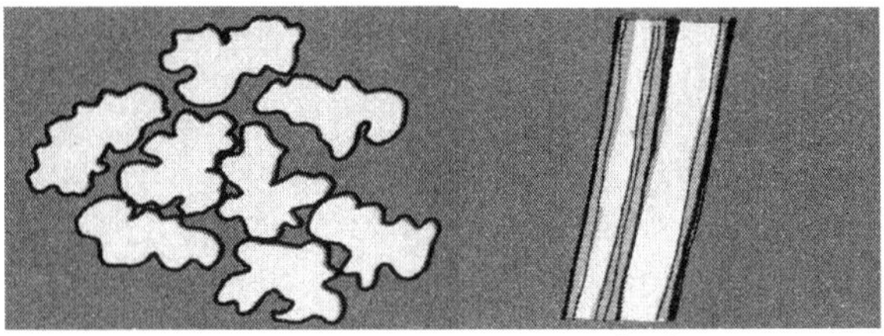

Figure 9.13 Microscopic view of acetate rayon fibre.

9.1.1.9 Triacetate

Triacetate is very similar in microscopic appearance to acetate (see Fig. 9.14). It may have clear striations and the cross-section is both lobed and somewhat serrated. Positive separation of triacetate and secondary acetate by microscopy is not satisfactory.

Figure 9.14 Microscopic view of triacetate.

9.1.1.10 Polyamide

Nylon is a manufactured fibre and, therefore, the diameter and the length of the filaments or staple fibre is determined by the producer and the ultimate end-use. Nylon is a transparent fibre when produced and can be made bright or dull. Nylon fibres are smooth and shiny. When viewed in cross-section, nylon is usually perfectly round (Fig. 9.15). Exception to this is trilobal nylons, e.g., Nylon 501, Autren, etc. Longitudinal magnification shows fibres of uniform diameter with a slight speckled appearance and which are relatively transparent.

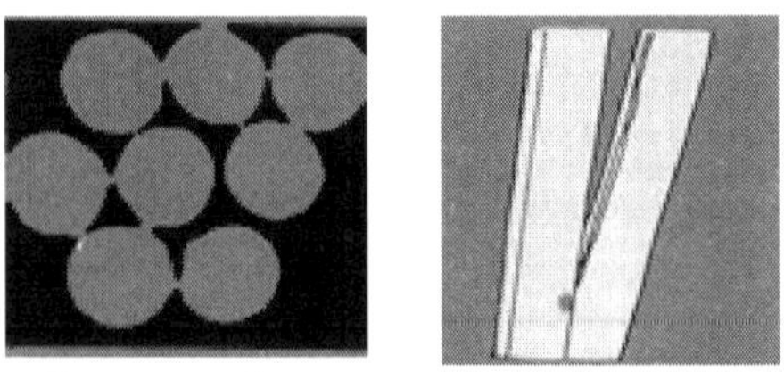

Figure 9.15 Microscopic view of polyamide fibres.

9.1.1.11 Polyester

A longitudinal view of polyester fibre exhibits uniform diameter, smooth surface and rod like appearance. The cross-section is usually round (Fig. 9.16). Variations are encountered in trilobal polyester, e.g., Dacron T-62. Fibres with dull and semi-dull appearance due to inclusion of pigment in the melt solution have a speckled effect due to changes in light reflection caused by the presence of the pigment.

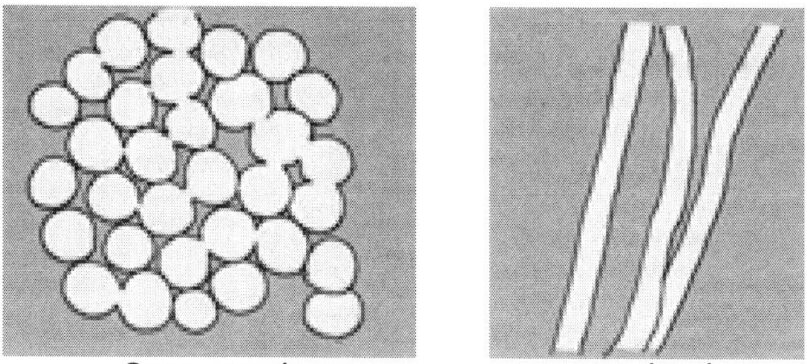

Cross section Longitudinal

Figure 9.16 Microscopic view of polyester fibres.

9.1.1.12 Acrylic

Cross-section microscopic views of acrylic fibres exhibit minor differences among the various kinds. Orlon possesses a dumb bell type shape (Fig. 9:17). Creslon and Zafran fibres are nearly round. Fine acrilan fibres have nearly round cross sections while fibres of high denier tend to be somewhat bean-shaped. Under the longitudinal view, fibre shows uniform diameter, rod like appearance, and some irregularly spaced striations or parallel lines. Those that have been delustered show the characteristic spotted effect caused by the pigment breaking up.

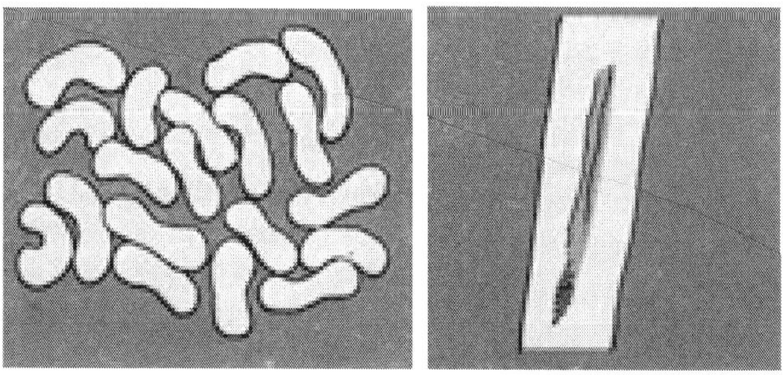

Cross section Longitudinal

Figure 9.17 Microscopic view of acrylic fibres.

9.2 Chemical tests

In addition to the microscopic test, chemical tests provide positive means of identifying the fibres. But chemical tests are not intended for the general consumer. However, it is a fact that some retail stores have their own laboratories for testing fabrics they procure.

9.2.1 Distinguishing animal from vegetable fibres with an alkali

As strong alkalies destroy animal substances, a 5% solution of Caustic Soda (NaOH) can be used to eliminate wool and silk fibres from a sample that contains a mixture of fibres. The action of this chemical is hastened by boiling the solution before the sample is immersed. The wool and silk fibres will be completely dissolved. The vegetable fibres will not be affected. Another test is necessary to identify those fibres.

9.2.2 Distinguishing vegetable and animal fibres with an acid

As dilute acids destroy vegetable fibres, 2% solution of sulphuric acid (H_2SO_4) may be used to distinguish vegetable from animal fibres. A drop of the solution is placed on a sample of the fabric, which is then placed between two blotters and pressed with a hot iron. The spotted area will become charred if the fabric is cotton, linen or rayon.

9.2.3 Distinguishing silk from wool fibres

Concentrated hydrochloric acid (HCl) dissolves silk completely and causes the wool fibre to swell.

9.2.4 Distinguishing nylon from other fibres

Nylon can be distinguished from acetate by attempting to dissolve it in acetone. Nylon will not dissolve in acetone. If the yarn or fabric is thought to contain wool and/or nylon, the fabric may be immersed in a boiling solution of sodium hydroxide. Nylon is insoluble in such a solution, but wool will dissolve. In fact, nylon's insolubility in boiling NaOH makes it easily distinguishable from most other fibres. The only common solvents in which nylon is soluble are concentrated formic acid, phenol and cresol.

9.2.5 Distinguishing polyester from other fibres

Like acetate and nylon, polyester is soluble in metacresol however, unlike acetate, it is not soluble in acetone; and unlike nylon, polyester is not soluble in concentrated formic acid.

9.2.6 Distinguishing viscose rayon, cuprammonium rayon and acetate rayon fibres

A solution containing equal parts of concentrated sulphuric acid and iodine (Crystal) may be used to distinguish each of the different fibres. When samples are immersed in this solution, a dark-blue colouration indicates viscose rayon; a light-blue colouration indicates cuprammonium, a yellow colouration indicates acetate.

Another test to distinguish acetate rayon from viscose rayon, when both the types are known to be used in a fabric, is to immerse the sample in a 50% solution of acetic acid. The acetate rayon dissolves; viscose rayon is not affected

9.2.7 Distinguishing acetate rayon from other fibres

If a sample of acetate rayon is immersed in acetone, the acetate fibres dissolve completely.

9.2.8 Distinguishing linen from cotton

Before any chemical test is done to distinguish linen from cotton, all surface dressings should be removed from the samples by boiling them for a few minutes in a solution of dilute sodium hydroxide (about ½%). Then, the following tests can be made.

Cotton fibres are weakened more than linen fibres if samples are immersed in concentrated sulphuric acid for two minutes. After samples are reused and immersed in weak ammonia, the cotton fibre dissolves, leaving the linen fibres unaffected. When cotton and linen samples are unused in a solution of iodine and zinc chloride, cotton is stained reddish purple and linen is stained blue to purple. When samples are immersed in a caustic soda solution, cotton remains white and linen turns yellowish.

9.2.9 Distinguishing arnel-triacetate from other fibres

When a specimen of fabric or blend of fibres is immersed at room temperature for one hour in a mixture of 90% methylene chloride and 10% ethanol, arnel and acetate will be dissolved, but other commonly used fibres will be virtually unaffected. Arnel can be differentiated from acetate by immersing a sample of fabric or fibres in benzoyl alcohal at 50 °C for one hour. Arnel is virtually unaffected but the acetate completely dissolves.

9.2.10 Distinguishing orlon-acrylic fibre from other fibres

Orlon is often blended with wool, cotton, silk, acetate rayon, nylon and polyester. Even when not blended, orlon may be mistaken for one of the other fi bres. If the fibre is orlon, it will not be affected by these common solvents; glacial acetic acid, chloroform, acetone and formic acid (88%).

9.3 Burning test

Burning test is a simple method to identify fibres, because what we need is simply a flame and the knowledge of burning properties.

Burning test is useful to all those who have no other method of identification. It is also useful as a confirmatory test, e.g., under microscope, regular nylon and polyester have similar appearance. It becomes difficult to identify them solely by using microscopic test; it is usually possible to distinguish between these two fibres by burning them. By this test, it is possible to identify the fibre as belonging to those classes which are outlined in Table 9.1.

Table 9.1 Burning test for some important fibres

Fibre	Flame characteristics	Odour	Ash
Cotton	Large yellow flame burns quickly	Burnt paper smell	Light, feathery grayish, black as denote mercerized cotton
Wool	Small slow flickering flame, sizzles and curbs ceases flaming when withdrawn from flame	Strong animal stench such as burning feathers or hair	Crisp dark ash, irregular shape that can be crushed easily
Silk	Burns slowly, ceases flaming when withdrawn from flame	Gives off an animal odor but less pronounced than wool	Round, crisp, shiny black beads, easily crushed with fingers
Rayon	Ignites quickly like cotton, sometimes faster	Burnt paper smell	Light feathery black ash

Nylon	Melts and shrinks from flame before burning	Pungent	Hard round bead that cannot be crushed
Polyester	Melts before burning, burns slowly	Slightly sweetish	Brittle, beady ash
Linen	Like cotton, sometimes slow burning if yarns are coarser	Burnt paper smell	Like cotton, sometimes if yarn is thicker
Acetate	Flames quickly, material puckers, sputters, melts drips like tar	Acid such as vinegar	Brittle, black ash, difficult to crush between fingers
Acrylic	Melts, burns rapidly	Acid	Hard black bead
Fibre glass	Does not burn, melts at temperature over 1500° C		Melts

To recognise the composition of fibres in fabrics by the burning test, we should know how the yarn ignites when the flame is applied, the condition of burnt portion and the odor produced by burning.

9.3.1 Limitation of burning tests

Burning test may give a due to the type of fibre, whether it is natural or manmade, but it has certain limitations.

Burning behaviour of a fibre or fabric may be affected by the type of finish imparted to the fabric. For instance, flameretardant finish on cotton fabrics greatly reduces the degree of flammability whereas napping or brushing of the same fabric may increase the rate of bumıng.

Also some of the finishes imparted to yarns or fabric will mask odour from the fumes.

Use of fibre blends may also complicate the identification of fibres by burning. Two or three fibres burning together in one blended yarn or fabric may be difficult to distinguish.

Some fibres burn almost identically. For instance, cotton linen and viscose rayon show a large yellow flame, rapid burning, and produces light feathery ash that gives off an odour similar to that of burnt paper. In such cases, burning test is of no value in distinguishing these three fibres from one another. But this test is important in distinguishing cotton, linen and viscose rayon from other fibres.

Many manmade regenearated and synthetic fibres react immediately to approach of flame, e.g., both acetate and Arnel,burn to a brittle black ash, flame sputter, give off an odour of acetic acid.

Dynel, Verel and Saran do not support combustion; but they melt and leave a brittle and beady ash.

Nylon, Orlon, Acrilan and Cresian melt before burning and leave a hard round bead.

These limitations or similarities discussed provide a basic clue to fibre type in the preliminary examination, but falls short of identifying the individual fibres separately in majority of cases.

9.4 Feeling tests

Feeling test is a subjective test. This requires skilled perception if it is to be of any value on a shopping tour. Skilled perception is acquired only after handling many different fabrics over a period of time. To understand what is meant by to feel a fabric; place the finger on a sample of wool. The heat generated by the finger remains in the area because wool is a non conductor of heat. Consequently wool fabrics feel warm to touch.

When the finger is kept on a sample composed of the, vegetable fibres that is cotton, linen, or even rayon, the heat of the finger passes of because these fibres are conductors of heat. Therefore, they feel cold to touch.

Cotton is cool to touch and feels soft and inelastic. Linen is cold and smooth and has a leathery feel. Wool is warm to touch and feels elastic and springy. Silk feels warm, smooth and elastic.

9.4.1 Limitations of feeling tests

Manmade fibres cannot be detected by feeling only because spun rayon can be made to look and feel like cotton, wool, or linen. Spun nylon may look and feel like linen or cotton. Vicara fibre may look and feel like wool.

Some of the newer fibres may even look and feel like each other in fabric form, as in the case of acrylics which have slippery or silk feeling that the natural fibres do not possess.

As there are various developments in the finishing technology and the finishes given to a fabric, the feel of the fabric is changed, and it becomes difficult or impracticable to distiguish the fibres by feeling.

9.5 Breaking tests

Breaking test is a non-technical test, yet it affords a convenient means of identifying fabrics.

Recognition of a yarn by the breaking test requires observation of the fibre ends, when the yarn is broken apart. A piece of about 300 mm length of yarn is required for the test.

Breaking of cotton yarn will show fibre ends that are short, even fuzzy and brush like. There is a characteristic curl to the ends caused by the natural twist of the fibre. The inelasticity of cotton yarn causes the yarn to snap when pulled apart.

When it is suspected that a yarn is composed of linen fibre, a long piece of yarn is necessary for testing because long staple may be anticipated. The breaking pattern of a linen yarn shows long, straight, lustrous fibre ends that are pointed and uneven at the tips. The fibre ends never curl. They point outward with marked stiffness, indicating lack of flexibility. Wool yarn has got a characteristic fuzzy surface. When the yarn is pulled gently for the breaking test, it stretches easily because of its remarkable elasticity. The breaking pattern shows fibre ends that are wavy and spiral.

Silk yarn stretches easily when pulled because silk ranks close to wool in elasticity. The fibre ends, when broken, appear fine and lustrous.

Viscose rayon in dry condition does not break easily. But, when moistened, a slight pull breaks the yarn easily. When broken, the fibre ends create a pattern similar to that of a tree with numerous stiff branches.

9.5.1 Limitation of breaking tests

As spun rayon can easily simulate wool, linen or even cotton, its breaking pattern can be similar to the characteristic pattern of wool, cotton, or linen. When the burning test is used as additional identification, the rayon that resembles wool reveals the lack of animal odour which is the characteristic of wool. Rayon yarn, that resembles linen or cotton, can be detected by exerting tension on the yarn after moistening. The characteristic loss of strength that rayon shows when wet is evident.

It would be difficult to determine accurately whether a yarn is composed of nylon, cotton, orlan, dacron, or any of the other stronger synthetic fibres. These fibres are relatively strong and the degree of tension used in trying to break these yarns by hand cannot be measured accurately. Furthermore, when broken, the ends of these yarns may look like each other, or like wool, cotton or some fibre depending upon whether it is the filament, artificially crimped staple, or staple fibres that were used in the manufacture of the yarn.

Chemical structure of some textile fibres

Cellulose polymer	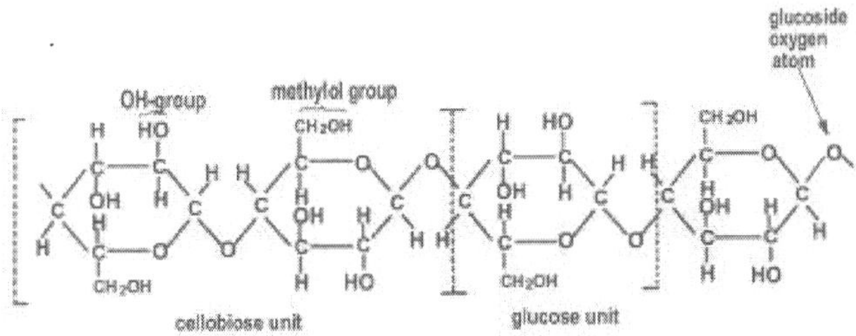
Cotton	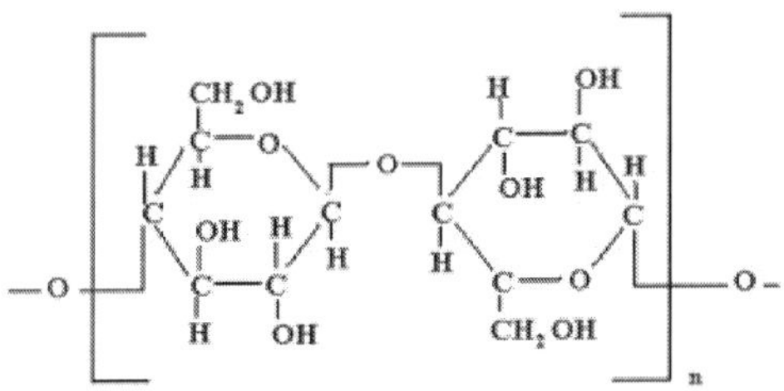
Silk	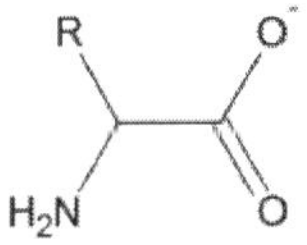
Fibroin protein in silk	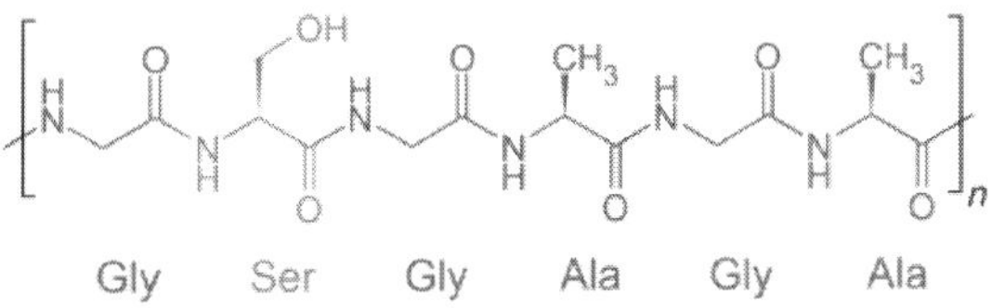

<table>
<tr><td>Wool</td><td>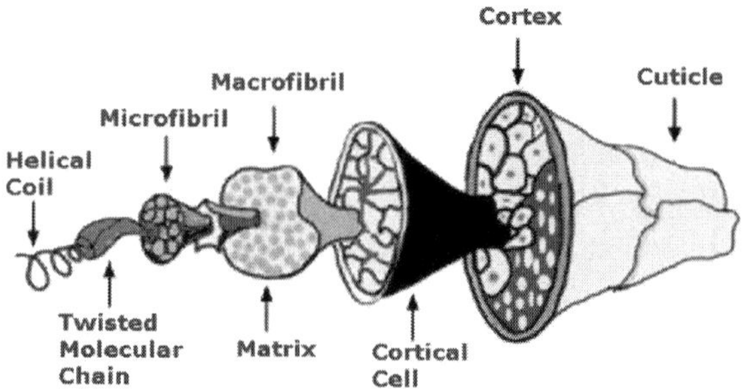

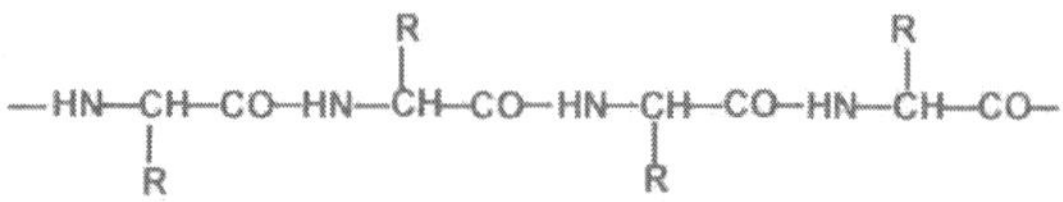</td></tr>
<tr><td>Kevlar</td><td>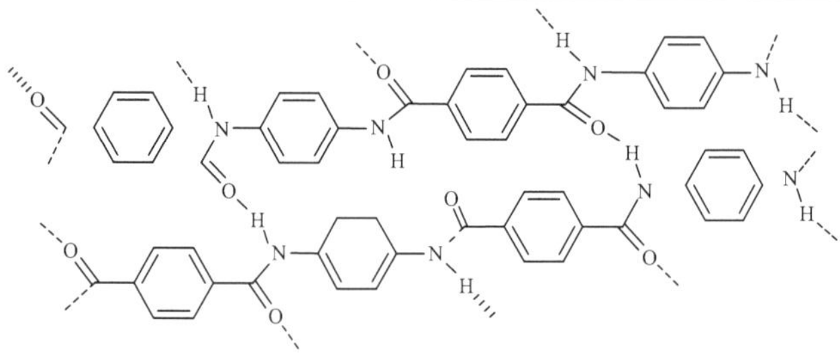</td></tr>
<tr><td>Cellulose acetate</td><td>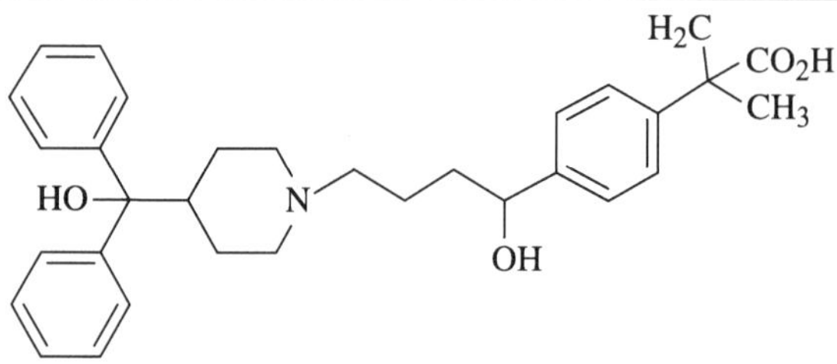
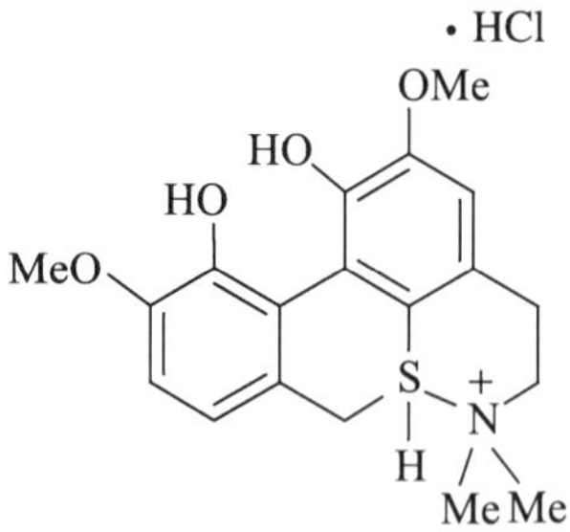</td></tr>
</table>

	Vinyl Acetate 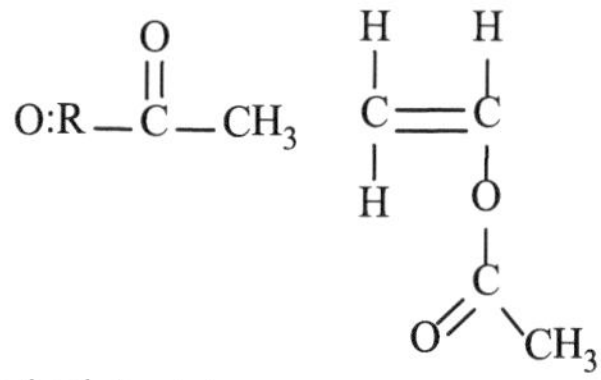 Polyvinyl Acetate (PVA) 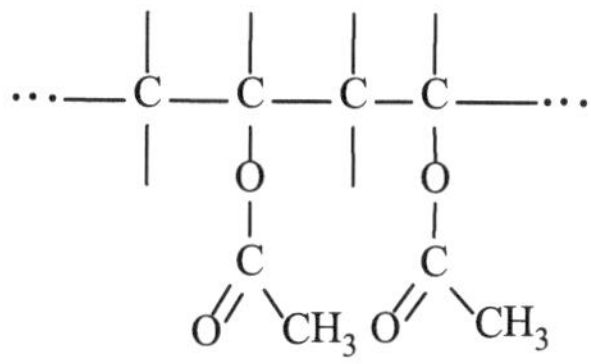
Cellulose triacetate	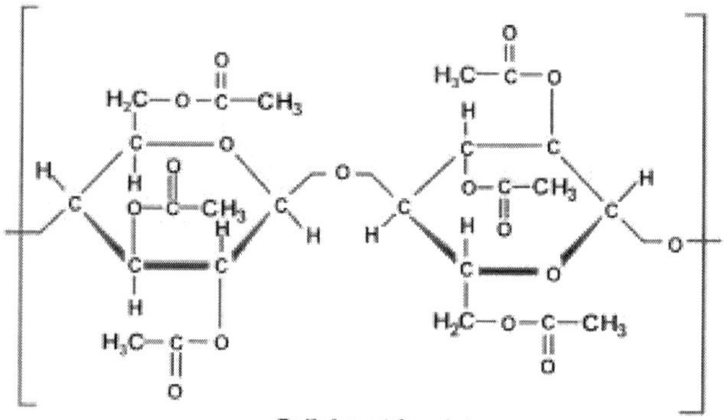 Cellulose triacetate
Nylon 6,6	$C_{12}H_{22}O_2N_2$ 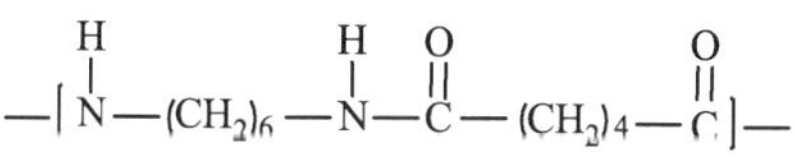 Nylon – 6,6
Nomex and Kevlar	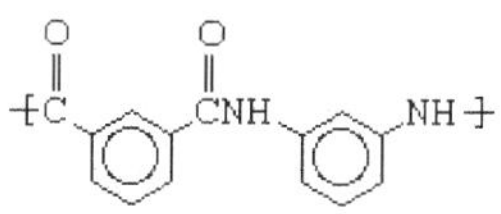 Nomex 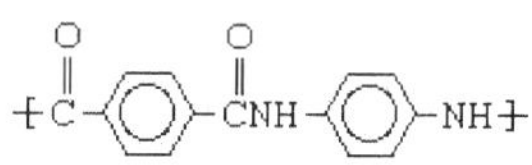 Kevlar
PET	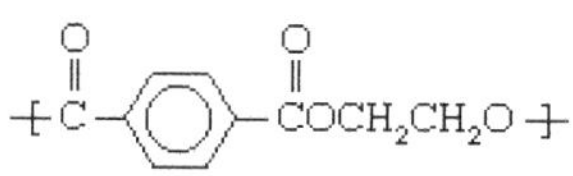

Polyolefin	$-[CH_2-CH(R)]-$ or $-[CH(H)-C(H)(R)]-$
Polyethylene	$-[CH_2-CH_2]-$
Polystyrene	$-[CH_2-CH(C_6H_5)]-$ or 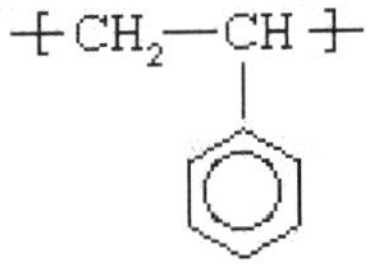
Polyvinyl chloride	$-[CH_2-CH(Cl)]-$
Polyacrylonitrile	$-[CH_2-CH(CN)]-$
Polymethyl methacrylate	$-[CH_2-C(CH_3)(CO_2CH_3)]-$
HEMA and Cyanoacrylate	$CH_2{=}C(CH_3)CO_2CH_2CH_2OH$ and $CH_2{=}C(CN)CO_2CH_3$
Polymethyl acrylate	$-[CH_2-CH(CO_2R)]-$
Polytetrafluoroethylene	$-[CF_2-CF_2]-$ or $-[CF_2-CF_2]-$

Aromatic AB polyamide: poly (p-benzamide)	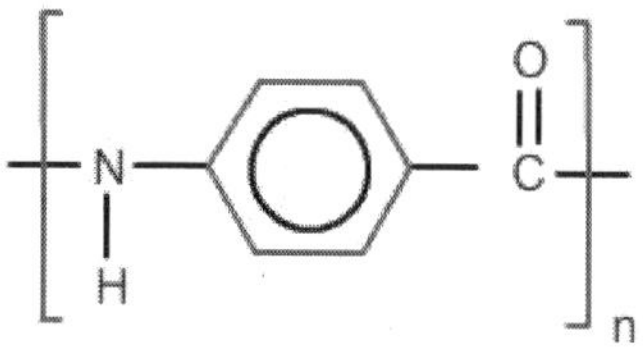
Aromatic AABB polyamide: poly (m-phenylene isophthalamide) MPDI	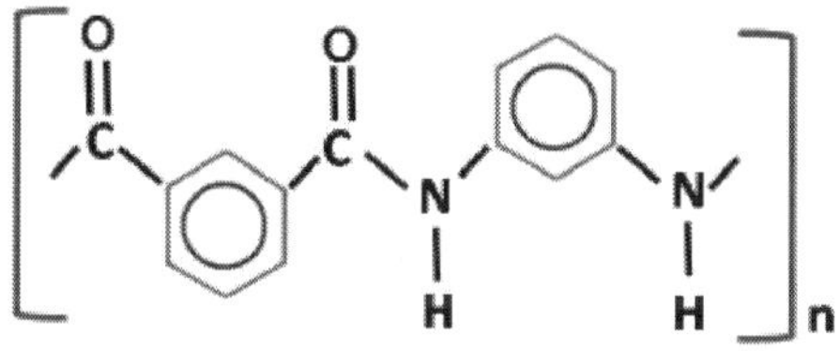
Poly(p-phenylene terephthalamide) PPTA or PPD-T	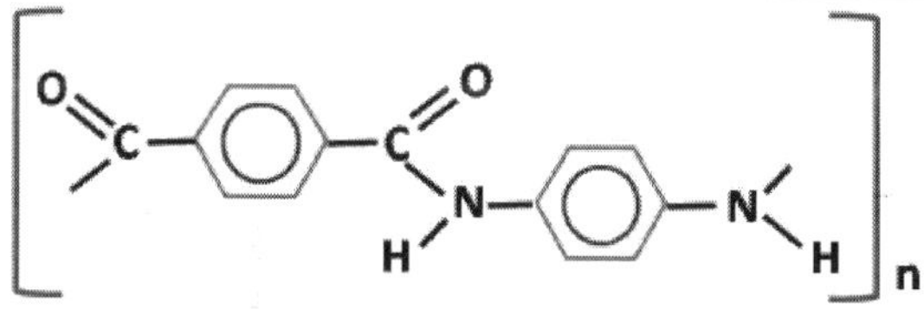

Appendix-II

End uses of some textile fibres

Fibre	Application
Vegetable fibres	
Cotton	Used universally. Major end uses include wearing apparel, home furnishings, medical-gauze, bandage, sanitary towels, cotton swabs, agro textiles, cosmetics, chemical industries, explosives, and so on.
Flax	Flax fibres are processed for use in carpet yarn, rope, geotextile (netting or matting), traditional carpets, paper, sacks, etc. They are also used in the non-woven, molding, and composite technology industries for the manufacturing of non-woven mats and carpets, composite boards as furniture materials, automobile door panels and headliners, etc. Special effects in lustre damasks, In India and America plants grown for seed (linseed).
Hemp	Paper, textiles, biodegradable plastics, health food, fuel, coarser textiles, sail-cloth, rope and twine, cordage of varying tensile strength, durable clothing and nutritional products.
Ramie	Coarse textiles. Cost of preparation for fine textiles prohibitive. Ramie is used to make such products as industrial sewing thread, packing materials, fishing nets, and filter cloths. It is also made into fabrics for household furnishings (upholstery, canvas) and clothing, frequently in blends with other textile fibers (for instance when used in admixture with wool, shrinkage is reported to be greatly reduced when compared with pure wool). Shorter fibres and waste are used in paper manufacture.

Kenaf	Rope, twine, coarse cloth (similar to that made from jute), and paper.
Hibiscus,	Coarse textiles. *H. Elams* has been extensively used in making mats.
Jute	Major end uses include bags, sacks (packaging), geotextiles (landfill covering, embankment reinforcement), hessian or burlap.
Sisal	Agricultural twine (binder twine and baler twine), general cordage, low-cost and specialty paper, dartboards, buffing cloth, filters, geotextiles, mattresses, carpets, handicrafts, wire rope cores, etc.
Sida	Coarse textiles; appears capable of substituting jute.
Lime or Linden	Matting and binder twine.
Mulberry	Paper and paper cloths.
Manila	Twine and ropes. Produces papers of special quality.
Sisal	Twines and ropes.
Phormium	Twine and ropes. Distinguished by high yield of fibre from green leaf.
Pine-apple	Textiles of remarkable fineness. Exceptional fineness of ultimate fibre.
Abaca	It is the strongest of the natural fibers. The fiber was originally used for making twines and ropes; now most is pulped and used in a variety of specialized paper products including tea bags, filter paper and banknotes. It is classified as a hard fiber, along with coir, henequen and sisal.
Banana	The outermost fibres of the shoots are the coarsest, and are suitable for tablecloths, whereas the softest innermost fibres are desirable for kimono and kamishimo. The banana fibre rugs are woven by the traditional Nepalese hand-knotted methods, and are sold Rug Mark certified.
Coir	Red coir is used in floor mats and doormats, brushes, mattresses, floor tiles and sacking, twines. Pads of curled brown coir fibre, made by needle-felting, are shaped and cut to fill mattresses and for use in erosion control on river banks and hillsides, insulation and packaging. A major proportion of brown coir pads are sprayed with rubber latex which bonds the fibres together to be used as upholstery padding for the automobile industry in Europe.

Animal fibres	
Wool	Major end uses include carpet, blankets, sweaters, coats, seat covers, clothing, horse rugs, saddle cloths, carpeting, felt, wool insulation and upholstery.
Angora wool	Angora wool is commonly used in apparel such as sweaters and suiting, knitting yarn and felting.
Silk	Major end uses include shirts, sarees, pyjamas, furnishing, decorative fabrics and parachutes. In early time silk was also used for making bulletproof vests.
Mineral fibres	
Asbestos	Various applications such as Chlorine Alkali diaphragm membranes used to make chlorine, Drywall and joint compound, Plaster, Gas-mask filters in pre 1960s, Mud and texture coats, Vinyl floor tiles, sheeting, adhesives, Roofing tars, felts, siding and shingles, "Transite" panels, siding, countertops and pipes, Popcorn ceilings, also known as acoustic ceilings, Fireproofing, Fire-resistant asbestos glove, Caulk, Industrial and marine gaskets, Brake pads and shoes, Stage curtains, Fire blankets, Interior fire doors, Fireproof clothing for firefighters, Thermal pipe insulation, Filters for removing fine particulates from chemicals, liquids and wine, Dental cast linings, HVAC flexible duct connectors, Drilling fluid additives, etc.
Regenerated	
Acetate	Linings, wedding and party attire. In home furnishings it is used for curtains, bedspreads, slip covers.
Viscose	It is used for shirts, dresses and linings. It has a soft handle, a good drape and can be washed and ironed. However, it has low warmth and is absorbent and slow-drying.
Synthetic fibres	
Nylon	Ropes, airbags, parachute, inflatable, tyre cords, dress materials, active sportswear, fleece jackets, socks, seat belts, carpets, nonwoven, webbing/luggage,
Polyester	All kinds of cloths, pillows, carpets, curtains, filters, for raincoats, fleece jackets, children's nightwear, medical textiles and working clothes.
Acrylic	In apparels it is used for making sweaters, socks, children wear, sportswear, carpets, industrial/apparel, hygienic/barrier cloth/bactericidal, nonwovens/paper, cement reinforcement, blankets, floor coverings and outdoor furnishings. Also used as a precursor for carbon fibre.

Polyethylene	Grocery bags, shampoo bottle, children toys and raincoats.
Polypropylene	Ropes, insulation for electric cable, sanitary products. Biaxially Oriented polypropylene (BOPP), clear bags, rugs and mats, Under armour clothing, roofing membranes, filter media in water filters, medical use is in the synthetic, non-absorbable suture prolene.
Kevlar	Bullet resistant vests, bullet resistant face masks, woven rope and in cable, and as inner lining for some bicycle tires.
Spandex–Lycra, Elastene	Garments, especially in skin-tight garments, active wear, athletic, aerobic and exercise apparel, belts, bra straps and side panels, competitive swimwear, cycling jerseys and shorts, dance belts, gloves, hosiery, leggings, netball bodysuits, orthopedic braces, rowing unisuits, cross-country race suits, ski pants, skinny jeans, slacks, miniskirts, socks and tights, swimsuits/bathing suits, underwear, wetsuits, zentai. Compression garments such as: foundation garments, motion capture suits, Shaped garments such as: bra cups, support hose, surgical hose, superhero outfits, women's volleyball shorts, wrestling singlets, Home furnishings, such as microbead pillows, etc.

Suggested further reading

1 J. E. Booth, *Principles of Textile Testing*, National Trade Press, London, 1961, Ch. 8, pp. 309-325.

2 A. Bayn, et. al., "Development of bandages using bamboo fibres" Asian Textile Journal, June 2008, pp. 74-77.

3 A. J. Hall, "Developments in Bi Component Fibres" Textile Weekly, January 1968, p. 12.

4 A. E. Haarer, "Jute - Substituted Fibres," Wheatland Journals, 1952.

5 A.R.Atkinson, *Jute - Fibre to Yarn*, B.I. Publications, Bombay, 1965.

6 B. Luriak, *The Identification of Textile Fibres*, Sir Issac Pitman & Sons Ltd., London, 1954.

7 B.C. Goswami, JG. Martindale, and F.L Scandino, *Textile Yarns Technology, Structure and Application*, Wiley Interscience, London, 1977, pp. 137-143.

8 H.V Sreenivasamurthy, - Impact of Manmade Fibres on Wearing Habits and Fashion by, 'Modern fibres', Vol. 5, No. I, 1974, p. 30.

9 C. H. C Cansdale, *Cocoon Silk*, Sir Isacc Pitman & Sons Ltd., London, 1937.

10 C.K. Chiang, et. al. - Phys. Rev. Lett. 39, 1098, (1977).

11 CAT Dikerson, et al - Naturally coloured cotton

12 Christopher Pastore and Paul Kiekens, "Surface characteristics of fibres and textiles," Part 1 and Part 2, Textile World.

13 Corolzzyn A. Farnfield and P.J Alvey, *Textile terms and definitions*, 7th edition, The Textile Institute, Manchester, USA, 1975.

14 D. S. Hamby, *The American Cotton Handbook*, Vol. 1, Interscience Pub. Inc., 1949.

15 E.P.G. Gohl and LD. Vilensky, *Textile Science an Explanation of Fibre Properties*, Longman Chashire Pvt. Ltd., 1980.

16 E.R. Post, M. Orth, P.R. Russo, N. Gershenfield, - IBM System. J. 39, 840 (2000).

17 Falkai B. V, "Production and properties of microfibres and microfilaments," ITJ Feb 1991, pp. 62-70.

18 Gopala Krishna and Nitya Kumar, "Organic cotton an overview," – Asian Textile Journal (ATJ), January 2008, pp. 35-42.

19 Gordon P. Bierwagen, "Major Industrial Polymers," Encyclopaedia Britannica.

20 H. D. Keiser, "Use of fibre glass by the army and navy, Mining Technology," vol. 7, No. 3, May 1943, p.7.

21 H.G. Howell, KW. Mieszkis, and D. Tabor, *Friction in Textiles*, Butterworths, London, 1959

22 Isabel B. Wingate, *Fairchilds Dictionary of Textiles*, 6th Edition, Pub: Universal Pub. Corporation, Mumbai, 1985.

23 J. C. Cook, *Handbook of Textile Fibres,* Vol. I & II England, 1968.

24 J.M. Preston, - Fibre Science, Manchester, USA, 1949.

25 J.W.S. Hearle, *High performance fibres*, Textile Institute and Woodhead Pub. Ltd., England, 2001.

26 J.W.S. Hearle and R.H. Peters, *Fibre Structure*, Butterworths, London, 1963.

27 J.W.S. Hearle and RH. Peters, Moisture in Textiles Butterworth Scientific Pub., Manchester USA, 1960.

28 J.W.S. Hearle and W.E. Morton, "Physical Properties of Textile Fibres", London 1962.

29 James M. Vreeland, "The revival of coloured cotton," Jr. Scientific American, 1999 Vol. 280, page 112.

30 K. P. Hess, "Textile Fibres and their uses", New York, 1954.

31 K. R. Salhotra, - Spinning of Manmades and Blends on Cotton System, Textile Association, India. 1983, p. 6-13.

32 Kamoogawa, Textile Weekly, June 1968, page 708.

33 K. D. Langley, "Quantitative Analysis of Fibre Blends," Blended Textiles, Ed. M.L Gulrajani, Textile Association (India) 1981, pp 210-215.

34 Kyle Ward, *Chemistry and Chemical Technology of Cotton*, Interscience Pub. Inc., 1955.

35 L Hunter, *Textile: Some Technical Information and Data II* SAWTRI Special Publication, July, 1978.

36 L.R.G. Treloar, *Introduction to Polymer Science*, Wykeham Publications Ltd., London, 1970.

37 M. L. Gulrajani, "Blended Textiles", Textile Association India, p. 94-109.

38 M.D. Bhavsar and H.B. Nagoria, "Staining Agents for Identification of Fibres", Manmade Textiles in India, June 1983, pp 273-275.

39 M.L. Rider and S. K. Stephenson, *Wool Growth*, Academic Press, London, 1908.

40 N. S. Kaplan - Textile fibres

41 N. S. Kaplan, - A practical guide to fibre science DK Agencies Pvt Ltd, DKLPV-4797 (HBD)

42 NIIR Board of consultants and Engineers, Natural Fibers Handbook with Cultivation & Uses, Asia Pacific Business Press Inc. ISBN 8186623981

43 Nonnan Hollen, Jane Saddler, and A.I. Langford, - Textiles', London 1979.

44 O. S. Hamby, - The American Cotton Handbook', Vol. 1, New York, 1965.

45 P A. Koch, - 'Microscopic and Chemical Testing of Fibres', Chapman and Hall, London 1963.

46 P. R. Rankilor, - Membranes in ground engineering, Wiley 1982.

47 Pal S. K., - Microfibre – production properties and application, textile asia, vol. 24, Jan 1993, page 53-58.

48 Paolo C. Arboni, - 'Silk-Biology, Chemistry and Technology', Chapman&Hall Ltd., WC-2, 1952.

49 Phyllis. G. Tonora, - 'Understanding Textiles'. London, 1982.

50 Premamoy Ghosh - Fibre Science and Technology, Tata McGraw Hill Pub. Co. Ltd., New Delhi, 2004

51 Purane, S. V., Panigrahi N. R - 'Microfibres, microfilaments and their applications' Autex research journal, Vol. 3, spet. 2007, page 148-158.

52 R. Jeffries - 'Bicomponent fibres' Merrow, Watford, England, 1971.

53 R. Meredith - 'The Mechanical Properties of Textile Fibres', Interscience Inc., NY 1956.

54 R. W, Moncrieff - 'Artificial Fibres,' STRAN W2, 1975 Ed.

55 R. W. Moncrieff - 'Wool Shrinkage and its Prevention', London, 1953.

56 R. W. Moncrieff - 'Manmade Fibres', London, 1970.

57 R.H. Baughman - Synthetic Met. 78, 339 (1996).

58 R.M.N Arib, S.M. Sapaun, M.M.H.M.Ahmad, H.M.D Khairul Zama - Mechanical properties of pineapple leaf fibre reinforced polypropylene composites, Materials and Design, Vol 27, Issue 5, 2006, page 391 to 396

59 S. Lam Po Tang, G.K. Stylios - Int. J. Clothing Sci. Technol. 18, 108 (2006).

60 S. Wagner, E. Bonderover, W.B. Jordan, J.C. Sturm - Int. J. High-Speed Electron. Sys. 12, 391 (2002).

61 S. Y. Nanal - High speed spinning of polyester and its blends with viscose, Wood Head publishing, India (P) Ltd. 2009.

62 V. A. Shenoy - 'Technology of Textile Processing, Vol. I, -Textile Fibres, Sevak Bombay-1977.

63 V. Parthsarathi - Nano Technology adds value to textile finishing – , Indian Textile Journal, Jan 2008, P57-61.

64 W. Krauss - 'Mathew's-Textile Fibres', Page 1092-1148Edited by H.R. Mauersbarger, London 1954.

65 W. V. Bergen - 'Wool Handbook', Vol. I, Textile Book Service, N, Jersey, 1963.

66 W.B. Achwal and H.J. Gore - 'Analysis of Blends Containing Cotton and Regenerated Cellulosic Fibres'. Journal of Textile Association, March 1974, page 13-16.

67 W.I. Garner - Textile Laboratory Manual Vol. 3 : Fibres', Heywood Books, London 1967.

68 W.T. Astbury - Fundamentals of fibre structure 1933.

69 Z.K. Walczak - Formation of Synthetic Fibres, Gordon and Breach Science Publishers, London, 1977.

70 Dictionary of Fibres & Textile Technology, Pub: Hoechst Celanage Corporation, 1975.

71 Classification of textile fibres - http://www.tessileesalute.it/flex/cm/pages/ServeBLOB.php/L/UK/IDPagina/192

72 FAO Agricultural Bulletin, 'Non.Mulbery Silks', FAO, UN-Rome, 1979.

73 Tecoya Trends – Textile Daily, April 16, 2009 issue

74 www.organicexchange.org

75 wikipedia.org

76 Growing organic Cotton some systematic approach - http://www.globecot.co.in/organic/new/organic_co_systematic_approach.html

77 Pineapple leaf fibre fabric dates back to hispanic times, March 12, 2009 (Philippines), Fibre2fashion News Desk – India

78 "Nanotechnologies for the Textile Market" (Cientifica, London, April 2006).

79 The impact of polyester blends, where do, wool cotton, rayon Stands now? Textil Month, 1969 Jan. P. 79-93.

80 Textile fibre end use survey. Textile Organon, Nov. 1970.

81 Terene Technical Information Booklet, ICI Trade literature.

82 British Standard Handbook No. 11, 1963, page 391-432.

83 "Handbook of Methods of Tests' CTRL-ICRA, Bombay, 1968 page 153-159.

84 www.bambootex.com

85 www.swicofil.com

86 H.M. Taylor 'Abrasion in Fabrics', Textiles, Vol. 7, June 78.

87 www.top5ofanything.com

Index

About the author

TAI Ratna Dr. H. V. Sreenivasa Murthy is a B. Sc. (Textiles) I Class, 3rd rank holder from Bangalore University; M. Tech. (Textile Technology) I Class distinction from Madras University; and Ph. D. from IIT Delhi. Besides he is also a post graduate certificate holder in Business Management from St. Xavier's College, Bombay. He has published more than 86 technical research and review papers in national and international journals, reviewed five technical books and written many special reports on national and international conferences, seminars and exhibitions. He is the recipient of "Century Mills Best Technical Book Award" for the books; *Introduction to Textile Fibres* in the year 1989, and *Two for One Technology and Technique for Spun Yarn* in the year 1993. He is the fellow of the Textile Association (India) – FTA and fellow of the Institution of Engineers' (India) – FIE. He is also a Practicing Chartered Engineer and Textile Consultant. He is on the selection bodies and Board of Studies in UG and PG studies of various universities. He was visiting faculty member of National Center of Quality Management (NCQM), Mumbai and successfully conducted number of ISO-9000 awareness programmes in mills. He had successfully organized many short term courses for spinning executives and edited books on 1. *Material Management in Textiles*, 2. Success story of export oriented units, 3. Value engineering in textiles (special issue JTA), etc. He is the Chairman of the Professional Award Committee of the Textile Association from 1985 till today. He is instrumental in introducing fi veyear Graduate Member of Textile Association – GMTA Course and three year Diploma level Associateship courses. He had successfully completed a joint project on 'Feasibility of Processing Indian Cotton and Blends in Air-jet spinning' along with KYOTO INSTITUTE OF TECHNOLOGY. Faculty of fi bre science, matsugasaki, Sakyo-ku-Kyoto, JAPAN and Central Institute of Research on Cotton – CIRCOT, Bombay. He has the working experience of two years in industry before joining Wool Research Association (WRA), Bombay then joined Veermata Jijabai (Victoria Jubilee) Technical Institute (VJTI) Matunga, Mumbai in September 1971. He has retired from VJTI as Professor and Head of the Department of Textile Manufacture in the year 2003. He has more than 32 years teaching and research experience.

From April 2005-2010, he was working as "Advisor" to Department of Textile Technology at Shri Vaishnav Institute of Technology & Science, Indore. From March 2007-12, he was working as 'Consultant' to NMMIS-MPSTME, developed from scratch Centre for Textile Functions, Shirpur (Maharashtra) - a Textile Diploma course, mainly focusing on Mentoring in the industry. From March 2009 to 2015, he was on the Maharashtra State Board of Technical Education- MSBTE, as chairman, Textile engineering group.

He has widely traveled abroad on one month study tour to MILAN – Italy through TECOYA initiative Australia on IWS and Australian Wool Federation invitation, and SOTSE's invitation to JAPAN and South Korea. He has also visited Singapore and Germany and other Asian countries.

He is recipient of Service Gold Medal and Service memento of the Textile Association (India). He was awarded TAI Ratna title on the Platinum Jubilee celebration of The Textile Association (India) on 9th April, 2014.

He is connected with various social and cultural organizations and has interest in literary activities, social services and yoga.